Qanat Knowledge

Ali Asghar Semsar Yazdi • Majid Labbaf Khaneiki

Qanat Knowledge

Construction and Maintenance

Springer

Ali Asghar Semsar Yazdi
International Center on Qanats and Historic Hydraulic Structures (UNESCO ICQHS)
Yazd, Iran

Majid Labbaf Khaneiki
International Center on Qanats and Historic Hydraulic Structures (UNESCO ICQHS)
Yazd, Iran

ISBN 978-94-024-0955-0 ISBN 978-94-024-0957-4 (eBook)
DOI 10.1007/978-94-024-0957-4

Library of Congress Control Number: 2016959456

Printed on acid-free paper

This Springer imprint is published by Springer Nature
The registered company is Springer Science+Business Media B.V.
The registered company address is: Van Godewijckstraat 30, 3311 GX Dordrecht, The Netherlands

Fig. 1 Two intersecting qanats near the town of Meybod, Yazd Province, Iran (Photo by Mehdi Khebredast)

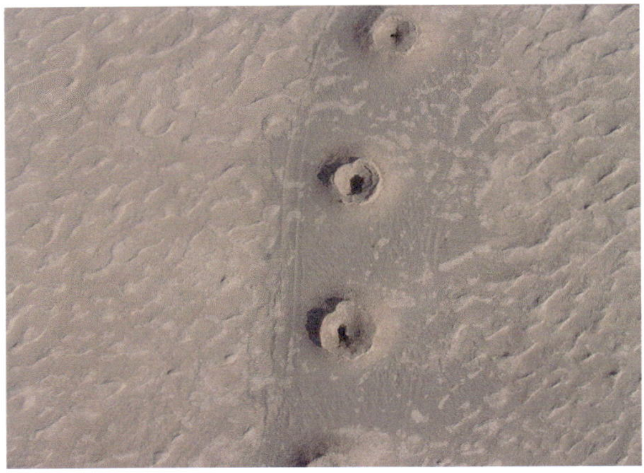

Fig. 2 Qanat near the town of Meybod, Yazd Province, Iran. Same photo as Fig. 1 from a different angle (Photo by Mehdi Khebredast)

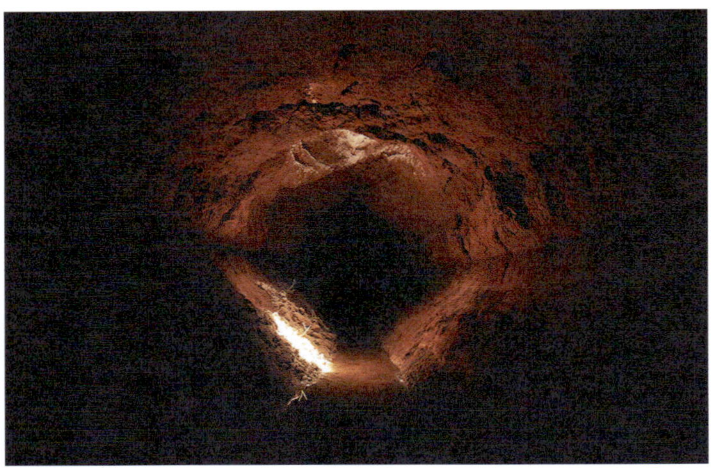

Fig. 3 A view from inside the qanat of Zarch in Yazd city, the longest qanat of Iran (Photo by Arshiya Khosravi)

Foreword

Qanats offer an ingenious solution for water supply in arid regions by creating a flowing stream of water where nature has not so provided. This book offers a ready solution for those who wish to learn more about this fascinating part of our water history and makes accessible to the wider world the traditional knowledge gained from building and maintaining qanats for more than 2500 years. There is much more here than a summary of the nature and distribution of qanats and a more extensive journey through the philosophy, methods, tools, and terminology of qanat design and digging than previously assembled. This most comprehensive book on the traditional engineering, construction, and maintenance of qanats is advised by the authors' experiences with qanat scholarship; their affiliation with UNESCO's International Center on Qanats and Historic Hydraulic Structures (ICQHS) and the affiliated Qanat College in Yazd, Iran; and their many years of practical experience in the field with indigenous practitioners of the art and engineering of qanats.

The great contribution of this work is the story it tells of the ingenuity and practical skills of the qanat masters who for centuries and generations have cut an uncountable number of tunnels through bedrock and alluvium using hand tools and homespun solutions to problems that would vex the most experienced university-trained engineers. Where does one begin to dig to ensure that the qanat tunnel will flow with water? How are practical considerations of landscape factored into the design? How are water quality and discharge measured? How does excavation proceed through bedrock and unconsolidated soil, and how is this knowledge of geology and pedology acquired? How are vertical wells and tunnels excavated to maintain proper air supply, light, and water flow? How does one deal with special problems like tunnel collapse, the accumulation of gasses and vapors, and the pooling of water during construction? How are tools and gauges designed, maintained, and used? How have qanats been incorporated into other structures like watermills, reservoirs, ice houses, and irrigation networks? And how are qanats cleaned, extended, maintained through the ages, and incorporated into modern water supplies?

Engineers will be amazed by the seemingly simple solutions to mining for water and the tool kit and skill set that developed them. Scholars will be fascinated by the practical wisdom of the indigenous qanat masters and the social organization and acquisition of environmental knowledge needed to develop and maintain qanats for so many centuries. The authors have assembled a how-to guide for the design, excavation, and maintenance of qanats. But, much more than that, we can also feel the resolve and instinct of the qanat master and sense the faces and hands that have built these great works.

Department of Geography
Oklahoma State University,
Stillwater, OK, USA

Dale Lightfoot

Preface

When the Iranian plateau entered a dry period some 4000 years ago and its aridness culminated later in the wake of a climate change, the ancient inhabitants turned to a genius technique which enabled them to stay for thousands of years. In the desert areas of Iran, the best arable soil is situated in the low lands, while the suitable water resources are found at higher altitudes at the base of mountain ranges where a relatively good precipitation takes place. In fact, the qanat system solves this contradiction and makes a link between the two curtail factors of agricultural production: soil and water. Digging through tens of kilometers of sediments with a variety of chemical and physical peculiarities entails a vast environmental knowledge which has been handed down from father to son over the past centuries. The ancient qanat builders had to come up with practical solutions for whatever problem they came across, the problems that related to different fields of science as we have determined today. Therefore, qanat is not only a simple gallery that extracts groundwater, but it used to put humans in closer contact with their surrounding nature and enhance their knowledge about their environment. The Neolithic livelihood was not sufficient to adjust humans to the arid lands where fresh water and arable soil were no longer at their disposal at the same place, but humans came to terms with such a harsh condition by inventing qanat, an underground tunnel which tapped groundwater and took it down to the less elevated lands. Qanat provided the inhabitants with the required skill which guaranteed their survival in one of the driest deserts of the world. In this book we have tried to shed light on the knowledge that humans have built up over centuries in order to construct or maintain qanats, and the figures and sketches are aimed to help us describe what is hidden underground.

Yazd, Iran

Ali Asghar Semsar Yazdi
Majid Labbaf Khaneiki

Acknowledgment

We wish to acknowledge all the qanat masters whose expertise and knowledge paved the way for this book to come about. All the information presented in this book has arisen from numerous interviews we have had with them during the past 10 years.

Also, special thanks go to Ms. Charlotte Kende and Mr. Mostafa Shafiee Kadkani whose illustrations flesh the book out and make it much easier for us to describe the qanat skills which are hidden underground. We also thank the International Center on Qanats and Historic Hydraulic Structures (UNESCO-ICQHS) and Tamadon Karizi Consulting Engineers (TKCE) for their precious cooperation and generous support. Eventually, we express our heartfelt gratitude to all those who have helped us in a way to shed light on the engineering feat that the traditional qanat masters and workers have created in the darkness of qanat.

Contents

About the Authors

Ali Asghar Semsar Yazdi was born in 1956 in Yazd, Iran. He finished his high school in Yazd and then moved to Shiraz to continue his study at the University of Shiraz in the field of mechanical engineering. In 1980, he graduated from the University of Shiraz and then was admitted to the Institute of Applied Sciences in Lyons, France. After he graduated from this institute with honors with a PhD degree in civil engineering, he returned to Iran and took part in founding such water organizations as Yazd Regional Water Authority, Yazd Water Museum, National Cloud Seeding Research Center, Qanat Collage, and International Center on Qanats and Historic Hydraulic Structures (UNESCO-ICQHS). He has published 55 scientific papers on water management and qanat system. He has also authored or co-authored the books *Qanats of Qasem Abad and Emamiyeh in Mashhad* (2016), *Qanat Tourism* (2015), *Qanat of Zarch* (2014), *Veins of Desert* (2010), *Qanat in its Cradle* (2012), *Qanats of Bam from Technical and Engineering Point of View* (2005) and *Qanat from Practitioners' Point of View* (2004), out of which the book *Qanat of Zarch* has won Iran's Book of the Year Award in 2016. In 2000, he initiated and held the first international conference on qanat, and in 2003, he was awarded the title "best researcher" in the Iranian Ministry of Energy. In 2012, he organized the international conference on Traditional Knowledge for Water Resources Management. From 2006 to 2013, he has been the director of the International Center on Qanats and Historic Hydraulic Structures which is a UNESCO category two center. From 2013 on he has worked for UNESCO-ICQHS as senior advisor.

Majid Labbaf Khaneiki was born in 1975 in Mashhad, northeast of Iran. He received his bachelor's degree in agricultural engineering, and then he graduated from the University of Tehran with a master's in geography and regional tourism planning. Now (2017), he is a PhD candidate in human geography in the University of Tehran. From 1998 to 2005, he has worked as a researcher for the Amirkabir Research Institute and the Iranian Academic Center for Education, Culture and Research (ACECR). Since 2005, he has been working as senior expert for the International Center on Qanats and Historic Hydraulic Structures, a UNESCO category two center (UNESCO-ICQHS). He has authored or coauthored the books, *Qanats of Qasem Abad and Emamiyeh in Mashhad*, *Qanat Tourism*, *Veins of Desert*, *Qanat in its Cradle*, *Water Division Systems in Iran*, *Water and Irrigation Techniques in Ancient Iran*, *Qanats of Taft*, *Qanat of Gonabad as a Myth, Qanats of Bam from Technical and Engineering Point of View*, and *Insight into Tabaran* which won Iran's Book of the Year Award in 2015. Tens of his articles have appeared in *Urban Tourism Journal (University of Tehran), Geographical Researches Journal, Mashhad University Journal*, etc. He currently works for UNESCO-ICQHS as researcher and senior expert.

Chapter 1
Introduction

Abstract This chapter starts with the engineering importance of qanat and the concept of cooperation which is in fact inherent in the technical traits and mechanism of qanat. The other issues discussed in this chapter are the geographical diffusion of qanats across the world, the controversial theories of its origin, its role in forming and nourishing human civilizations in arid and semi arid zones and comparison between qanat and other historic hydraulic technologies. We do not want to immerse ourselves in the issue of qanat and civilization, but at least in this book we outline this issue which is so important that it warrants another study.

Keywords Qanat history • Climate change • Human adaptation • Geographical distribution • Qanat civilization

1.1 Introduction

Qanat is a gently sloping subterranean canal, which taps a water-bearing zone at a higher elevation than cultivated lands. In 2014 there have existed some 37,000 active qanats running all over Iran, discharging about seven billion cubic meters groundwater a year. At a glance Qanat is nothing but a horizontal tunnel which drains out groundwater, but digging this tunnel entails a variety of sciences and technologies. Qanat is a feat of technology left from our ancestors, a feat which is hidden underground, but its technical importance is obvious, not less valuable than such surface structures as bridges, castles, towers, etc. Qanat enjoys extended structures and sometimes its length reaches tens of kilometers. Qanat passes through geological formations and faces different conditions and obstacles, so the qanat masters' efforts to solve these problems led to the accumulation of knowledge in terms of Qanat construction over time, which has been handed down from generation to generation. Qanat is one of the most complicated traditional technologies which require knowledge on nature ranging from groundwater to management. This indigenous technology used to bring water efficiently from tens of kilometers away to the thirsty lands. In many of Qanats which were constructed hundreds of

A.A. Semsar Yazdi, M. Labbaf Khaneiki, *Qanat Knowledge*,
DOI 10.1007/978-94-024-0957-4_1

years ago, no technical errors took place, and even their present structure bears witness to this fact. This source of knowledge behind the system of Qanat can be ranked among the wonders of human civilizations.

Traditional qanat know-how encompasses a wide range of indigenous knowledge from geometry to botany. In fact if there would not be interaction and cooperation between different fields of science, no qanat could come into existence. At the beginning of construction of a qanat, the workers should know about the plant species in the desert and their relationship with groundwater in order to designate the best spot to dig the first well. To do so they should also be knowledgeable on geological stratification to better estimate the depth of the bed rock and the sediment thickness. As they further the qanat construction, more and more fields of science would be engaged in their work. For example in Yazd, Iran, when the qanat masters dig overhead from the tunnel ceiling up toward the earth surface they apply some interesting mathematical equations to minimize their error. During our field studies we realized that a qanat master knew of the above mentioned mathematical equations but he talked about them in a different language. When we tried to translate them into the conventional mathematical equations, we were amazed at the fact that an illiterate traditional qanat master was using Pythagorean theorem in practice. Cooperation between fields of science even goes farther where medical knowledge is applied to cure the injured or sick workers or prevent the potential threats, where astronomy is used to calculate time for qanat water division.

Qanat engineering is a collective knowledge and expertise built up through centuries of cooperation. In the regions where qanats are the cornerstone of the local economy, sense of cooperation is pretty high. It has something to do with the nature of qanat which is deeply contingent on cooperation between those who benefit. Cooperation is perceptible in qanat from its very beginning to its management over hundreds of years of its age. This cooperation also spreads to the other realms of social life and comes to turn into sort of cultural genetics. That is why in central Iran local people can easily reach consensus over things apart from qanat. Cooperation is one of the main characteristics of qanat civilization, a social trait which is anchored in the technical and managerial quality of qanat.

Human is a social species whose development is deeply indebted to the sense of cooperation. When humans could invent language at prehistoric times, cooperation was put on a new track which led to the accumulation of knowledge and the genesis of great civilizations. Every social activity served as a core around which experiences and knowledge of the participants could build up. For example if the cavemen did not have to run after games in their primitive groups, they might not be able to mobilize their experiences in order to organize more complicated communities. Qanat system could play such a role in the communities of the arid and semi arid lands where scarcity or even lack of surface water drove people to search for other water sources underground through qanat technique. Technologies related to qanat are of great value, but we believe that the subtle role of qanat in creating and strengthening the sense of cooperation in local communities is not less valuable.

Qanat is a geographically extended system being sometimes tens of kilometers long. In fact a qanat gallery cuts through a variety of geological formations, runs beneath many human settlements and cultivated lands and is subject to numerous potential threats. Therefore a qanat cannot be built, maintained and operated just by an individual but it demands a vast cooperation from the community. This cooperation takes place from the very first moment of its construction to the last drop of water coming out from its gallery. To better get a handle on this issue, we can classify the fields of cooperation in qanat as construction-maintenance, operation, and technology transfer.

As mentioned, qanat is a long tunnel whose construction consumes a lot of time, energy and money. It is very difficult for an individual to afford the money required to build or even repair a typical qanat. Therefore the first step to build or repair a qanat is to bring together all the beneficiaries and pool their money in order to cover the qanat expenses. Qanat cost is not limited to the money spent on its construction, but it continues into its operation, because a typical qanat needs repair and cleaning once in a while in order to keep running. Therefore an elaborate financial system has evolved in terms of qanat over time. In some areas one may see some traditions and customs which are subtly associated with financial affairs of qanat. For example in some regions of central Iran the locals used to hold a wedding ceremony for their qanat when it came to dry up. They asked a widow in the village to become the wife of qanat and dwell in a house built just close to the qanat exit. The villagers treated with this wedding and the qanat's wife the same as they did with a real bride, and all the wedding customs were observed exactly like a real one. One of those customs was the wife's allowance which should be paid by any husband regularly in their culture. Since this weird husband was not able to fulfill its commitment, the village people raised an amount of money once in a while to pay to the bride on her husband's behalf. Of course the money collected this way was much more than what the qanat's wife really needed, so the rest went to qanat maintenance and cleaning. Thus those people were right in their belief that marrying off their qanat can increase its water. Although this custom has long been abolished, there is still its vestige in many Iranian villages. The allowance paid to a wife is called "Nafaghe" in Persian language but this term also now refers to the money that villagers collect to clean and repair their qanat. The name of that ancient custom is living on, though there is no longer a trace of it in reality (Fig. 1.1).

Cooperation is more apparent in the working team while a qanat is being built or refurbished. A working team consists of at least four workers which are closely connected like the cogs of a machine. The team head is the qanat master (muqqani) who is more experienced than the other workers and digs the tunnel end into the soil. A worker sits behind the master and collects the excavated soil and puts it in the buckets which are carried by another worker along the tunnel and hooked to a rope hanging from the nearest well. Finally a worker operates a windlass on the ground to coil up the rope and pull up the debris bucket and dump it around the well opening. These workers perfectly team up to make the most of their energy and time. Sometimes there are additional teams working simultaneously to speed up the qanat construction, and all of them are in a perfect harmony and coordination (Fig. 1.2).

Fig. 1.1 Traditional wedding ceremony for qanat

Fig. 1.2 A qanat working team with different specialized jobs

Cooperation is not limited to qanat construction or maintenance, but it becomes more visible when qanat is being used. As mentioned in most cases many shareholders together are entitled to a qanat whose water should be divided among them as fairly as possible. Sometimes over one thousand shareholders came to an understanding to select someone from among themselves to be responsible for water division. He should record the shares and keep track on any change and transaction in the existing shares, and oversee the water division to prevent any possible mistake. All the shareholders are fully in sync with the qanat water division system and everybody takes their shares according to a rotation of irrigation which allows them to irrigate their lands only at specified times. For example, if a farmer has an irrigation right of 2 h within a 6 day irrigation cycle or rotation, it means that he has the right to water his land just for 2 h once every 6 days. The duration of the irrigation cycle ranges from 6 to 21 days in all over Iran, but the average is between 6 and 16 days in most of the country, which is related to the cropping pattern. Sometimes all the shareholders decide to change the irrigation cycle to better adapt to likely climatic changes. If there would not be such a cooperation, no one could survive in the harsh environment of the Iranian desert. Fluctuation in the flow of a qanat can drive the shareholders to modify their irrigation cycle unanimously. If the discharge of a qanat decreases due to drought, the water shares would no longer suffice to irrigate the existing lands. For example, if a farmer has a water share of 2 h within a 6 day irrigation cycle, and the discharge of the qanat dropped from 100 to 50 l per second, he would no longer be able to water his entire land. To solve this problem he changes his crop to something more resistant to the dry condition, and then he receives a water share of 4 h once every 12 days not 6 days. By this means, the existing water can cover all his land, even though the qanat flow is low. This example applies to all the shareholders which are under the same circumstances. This coordination and cooperation could have ensured their survival in the face of water shortage in central plateau of Iran. The social structures of such region could have evolved this way to gain such a level of cooperation to better adapt to their environment.

Sometimes some shares of a qanat are devoted to public or religious purposes. This practice is called "Vaqf" whose revenue may be allocated to a mosque, a school, or helping orphans or paupers. Here cooperation protrudes from qanat itself and reaches out to the other parts of society. Social convergence reaches its peak where hundreds of qanat shareholders recognize a part of qanat water as public property and do not violate such social agreements.

Qanat also gives rise to cooperation in constructing and using some hydraulic structures associated with qanat. For example a water reservoir is filled up with the water coming from a qanat nearby. In the traditional towns and villages water reservoir plays a crucial role in supplying drinking water, which is deeply indebted to cooperation, because all the qanat shareholders should be unanimous to hand over their water to the reservoirs whenever needed (Fig. 1.3).

A watermill also operates with the qanat water. A watermill has an intricate relationship with the qanat, its shareholders and the local community from financial and social points of view. In a nutshell the right of building and running a watermill can be granted to an individual or a group of people but in fact the watermill belongs to the qanat whose water rotates the watermill turbine. Someone who is in charge of the watermill pays something in cash or in kind occasionally to the qanat owners, and then this payment is spent on the qanat rehabilitation and maintenance. All these things take place in a very systematic way in which all the factors are closely interwoven (Fig. 1.4).

Cooperation does not only occur in a particular qanat, but also it may take place between different qanats. In many valleys in Iran qanats usually run down the

Fig. 1.3 Water reservoir filled by qanat water

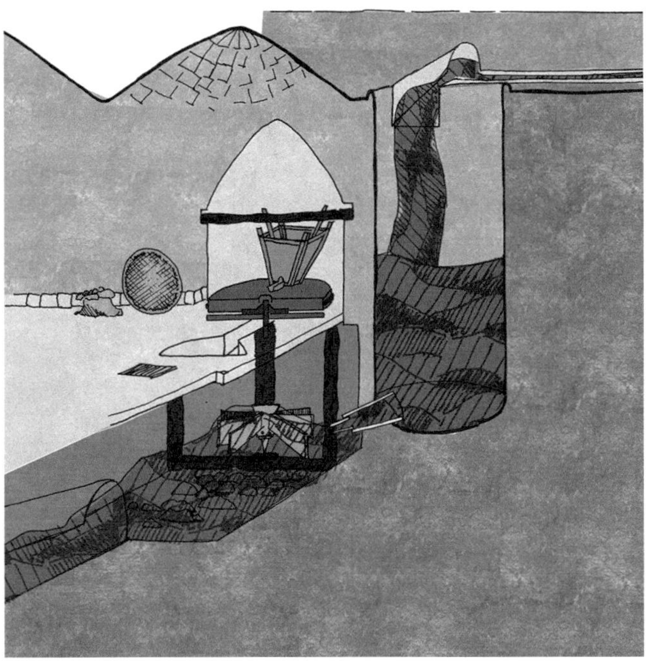

Fig. 1.4 Watermill operated by qanat water

mountain foots parallel to each other. These qanats originate from mountain foots overlooking a vast plain with thick layers of sediments. Precipitations on the mountains recharge the aquifer and seep into the qanats. In this case the owners of all qanats pull through to better maintain their qanats. If a qanat would be obstructed by a collapse, the qanat owners have the permission to dig a side tunnel just before

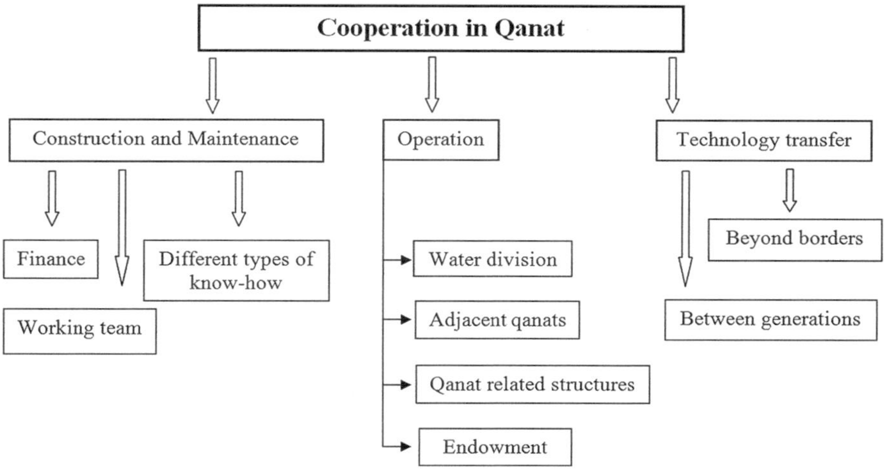

Fig. 1.5 Aspects of cooperation in qanat (*arrows* do not denote causal relationships)

the obstruction toward an adjacent parallel qanat. This water can be conveyed to their farmlands through their neighboring qanat, otherwise their lands would dry out and their crop would wither away. To do so, they measure the water flow before joining the neighboring qanat and then they receive the same amount downstream at the exit point. Diverting water to a neighboring qanat gives them enough time to fix the problem and remove the obstruction more comfortably. This cooperation benefits all the qanats in the valley whenever they are in trouble.

Qanat knowledge is in fact a historical process which has taken place through an interaction between nations and at same time between generations. From a historical perspective, qanat has been a vehicle for accumulation of humans' knowledge on their environment. This knowledge has been made up of various pieces coming from different nations and generations to portray a gorgeous picture of human coexistence with nature, embodying the concept of cooperation through both history and geography (Fig. 1.5).

1.2 Genesis of Qanat

It is Henry Goblot who explores the genesis of this technology for the first time (Goblot 1979). He argues in his book entitled "Qanat; a Technique for Obtaining Water" that during the early first millennium before Christ, for the first time some small tribal groups gradually began immigrating to the Iranian plateau where there was less precipitation than in the territories they came from. They came from somewhere with many surface streams, so their agricultural techniques required more water than was available in the Iranian plateau. So they had no option but to fasten their hopes on the rivers and springs that originated in the mountains. They faced two barriers; the first was the seasonal rivers which had no water during the dry and hot seasons. The second was the springs that drained shallow groundwater and fell dry during the hot season. But they noticed some permanent runoff flowing through the tunnels excavated by the Acadian miners who were in search of copper. These farmers established a relationship with the miners and asked them to dig more tunnels in order to supply more water. The miners accepted to do that, because there was no technical difficulty for them in constructing more canals. In this manner, the ancient Iranians made use of the water that the miners wished to get rid of it, and founded a basic system named qanat to supply the required water to their farm lands. According to Goblot, this innovation took place in Urartu[1] and later was introduced to the neighboring areas like the Zagros Mountains in Iran.

According to an inscription left by Sargon II the king of Assyria, In 714 BC he invaded the city of Uhlu northwest of Uroomiye lake that lay in the territory of the Urartu empire, and then he noticed that the occupied area enjoyed a very rich vegetation even though there was no river running across it. So he managed to discover the reason why the area could stay green, and realized that there were some qanats behind the matter. In fact it was Ursa the king of the region who had rescued the people from thirst and turned Uhlu into a prosperous and green land. Goblot believes that the influence of the Medians and Achaemenians made the technology of qanat spread from Urartu to all over the Iranian plateau.

Nevertheless some scholars go far back in antiquity regarding the qanats of Syria. They have maintained that the Hailan-Aleppo qanat, a 12-km long subterranean channel which functioned until the early part of this century, is coeval with the Aramaeans and their fortress at Aleppo (thirteenth century B.C.). However, this is highly unlikely as the best evidence (archaeological and written accounts) suggests that qanat irrigation was first invented in the Armenian-Persian region about 600–800 B.C. (Lightfoot 1997: 432–451). Jordanians today refer to qanats as “Roman canals”, or qanat Romani (kaneh Romani in northern Jordan). Most scholars believe that Jordan’s qanats were built by the Romans and used by the Byzantines from the first century B.C. to the seventh century A.D. (Lightfoot, 1996: 321–336).

The origin of qanat has caught the attention of many scholars, and there is some valuable research on this issue. For example, Haupt a German researcher has studied a qanat system in the area of Lake Van in Turkey (Haupt Lehmann 1925), and then Weisgerber conducted research on the history of qanats in Oman (Weisgerber Gerd 2003). Chauveau has studied qanats in the oasis of Kharagha’ in Egypt from archeological point of view (Chauveau Michel 2001), and Salvini has done some research on the historic hydraulic structures in Urartu (Salvini Mirjo 2001). Also, there are some scientific reports on the qanats of Oman and Iran, written by

---

[1] Strictly speaking Urartu is the Assyrian term for a geographical region, while the “Kingdom of Urartu” or the “Biainili lands” are the Iron Age state that arose in that region. That a distinction should be made between the geographical and the political entity was already pointed out by König. The landscape corresponds to the mountainous plateau between Asia Minor, Mesopotamia, and the Caucasus mountains, later known as the Armenian Highlands. The kingdom rose to power in the mid nineth century BC and was conquered by Media in the early sixth century BC. (http://en.wikipedia.org/wiki/Urartu).

Boucharlat, the French archeologist (Boucharlat Remy 2001: 170). Another scholar, Briant, has scrutinized the text of Polibius on the qanats of Iran (Briant Pierre 2001). An Iranian historian named Zohreh Cheraghi has provided a valuable anthology on the researches ever conducted on qanat history in her PhD thesis (Cheraghi Zohreh 2010).

Despite these studies that mostly search for a single inventor and try to pinpoint the origin of qanat, qanat is more likely to have come into existence in different places simultaneously and independently as a response to climate change in ancient times, though we do not rule out the geographical diffusion of qanat and exchange of technology between nations. For example, Goblot's theory may be valid in Urartu or neighboring regions, it is more likely that the first qanats were built in the central plateau of Iran at the mountain bases or along the valleys, though we do not deny the possibility of construction of qanats outside of this region for example in Oman or Urartu independently. It is hard to accept the theory that the qanat was first invented by the copper miners in Urartu and then introduced to the Iranian plateau and used by the farmers who lived some 1500 km away from its origin. Not only in the past, but also at present, the immediate reaction of any farmer is to dig into a spring when its water dwindles. In the mountainous region surrounding the Iranian desert there were many natural springs which supplied water to the small communities who lived there. In the wake of climate change, the precipitation reduced and accordingly many of the springs dried up or just trickled. In this situation the immediate reaction of the people might be to dig the same springs to trace the water, and after a while they ended up building a long tunnel with some shaft wells through which they could better haul the debris onto the surface. In fact, we consider that the natural springs led the people to construct the first qanats, and it is very likely that an ancient man would be inspired by a trickling spring to burrow back to get closer to the source of water (Labbaf Khaneiki 2015) (Fig. 1.6).

Probably that was how the system of qanat came into existence, probably in several regions simultaneously. Hasanalian who has conducted much research on Sialk[2] has come to the conclusion that the spring of Fin which once provided this ancient settlement with water was later manipulated and turned into a qanat (Hasanalian Davood 2006). Even a few years ago we witnessed this process in an off the beaten path village in southern Khorasan. In this village there was a natural spring with a discharge of about 3 l per second. After a drought broke out some 15 years ago, the water of this spring dramatically decreased and as a result the villagers made up their mind to deepen the spring to reach water again. They dug the spring horizontally up to 30 m and every year they extended this tunnel to keep the discharge steady. After 20 years that spring was turned into a qanat with two shaft wells.

[2] Sialk is a large ancient archeological site near the city of Kashan, in central Iran, tucked away in the suburbs, close to Fin Garden. The culture that inhabited this area has been linked to the Zayandeh Rud Civilization. At the site, there are actually two structures (necropoli) Sialk situated several 100 ft from each other. The three platforms of the larger ziggurat, however, still remain in place. Not much remains of the smaller structure. The Louver has also excavated a cemetery near the structures that has been dated as far back as 7500 years. (http://en.wikipedia.org/wiki/Tappeh_Sialk).

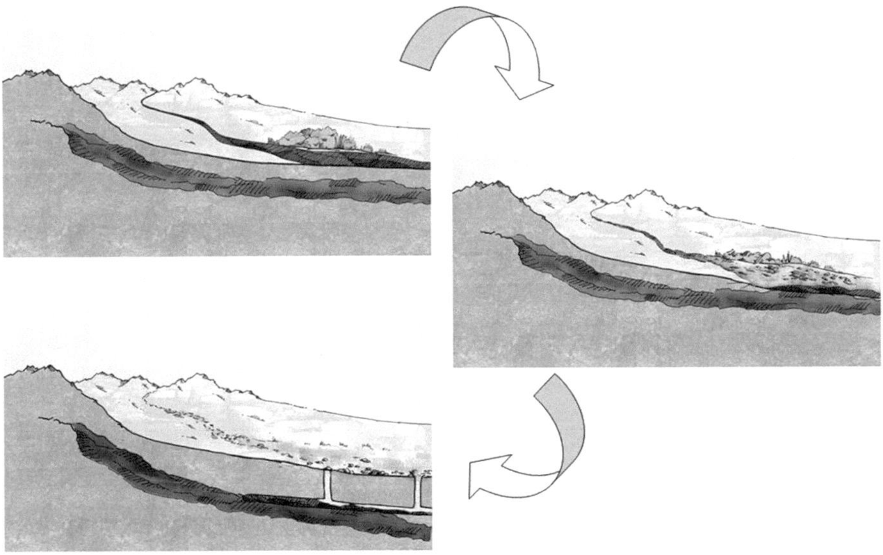

Fig. 1.6 Genesis of qanat as a response to climate change

This scenario can be seen beside the theory that the miners of Urartu invented the qanat as a byproduct, and could be repeated wherever enjoyed suitable conditions for qanats. Who knows how or even whether the farmers in the central Iran came in contact with those miners in Urartu and how they learnt this technique and how they brought it to the central Iran?

1.3 Geographical Distribution of Qanats

The arid and semi-arid regions of the world, whose rainfall shortage does not allow any permanent surface streams, but enjoying rich groundwater resources, have had a good potential to house the system of qanat. So, the system of qanat could be introduced and spread rapidly across such regions.

Iwao Kobori (Kobori 1964) believes that the qanat system, in all probability, developed in ancient Persia some 2500 years ago and then spread to Afghanistan and eventually along the Silk road as far east as China, as well as by Arabic cultures to the far west including Morocco and Spain.

According to Goblot (1979), qanats originated in the northwest of present Iran, dating back to 600–800 B.C. In 525 BC qanats were introduced to the southern coast of the Persian Gulf and then in 500 BC to Egypt, in 750 AD to Spain, in 850 AD to Southern Algeria, in 1520 AD to Mexico & Los Angeles, in 1540 AD to Pica in Chile and in 1780 AD to Turfan (NW China).

The system of the qanat spreads out between the latitudes 15° and 45°N, and this technique can be found even in such rainy regions as Germany. According to some studies, many of the countries between the abovementioned latitudes enjoyed the system of qanat and some of them still profit from this technique: for example, one can mention such Asian countries as Japan, China, Afghanistan, Pakistan (Baluchistan), Iraq (Kurdistan), Oman, Saudi Arabia, Iran, India, Syria and Azerbaijan (Nakhchivan).

Also, most of the North African countries like Egypt, Libya, Tunisia, Algeria, and Morocco enjoyed the benefit of qanats some of which are still active. In the Canary Islands a considerable number of qanats has been reported. In Europe including Spain, Greece, Germany and Sicily there existed qanats some of which were used as a subterranean canal just to convey water from a surface source to the desired place.

In the new world, the qanat has not played a vital role in supplying water, but some running qanats have been seen in Mexico (Parras), Peru (Nasca), Chile (Pica), Hawaii (Honolulu), USA (California and Los Angeles).

1.4 Water Division System in Qanat

Traditional water division means all the actions the local farmers take in order to regulate water division, irrigation related subjects and preservation of water resources. They have traditionally established some complicated systems in order to divide water among the farmers or the shareholders of a water resource, and irrigation rights are based on landownership or time shares within a certain period of rotation.

Water for irrigation is owned by shares. In fact the farmers take turns bringing qanat water to their land. For a particular shareholder, the interval between two irrigations means an irrigation cycle or rotation pattern of irrigation water called "Madar". For example, if a farmer has an irrigation right of 2 h within a 6 day irrigation cycle, it means that he has the right to water his land just for 2 h once every 6 days. The duration of the irrigation cycle ranges from 6 to 21 days in all over Iran, but the average is between 6 and 16 days in most of the country, which is associated with the cropping pattern and soil condition. In terms of wheat and barley which are the most common crops in Iran, the best interval between two irrigations is 12 days, and that is the length of the most common irrigation cycle (Semsar Yazdi and Labbaf Khaneiki 2010: 106).

Qanat water division system can match up with all likely changes in the volume of water during a year, while satisfying the farmers' irrigation needs. To measure the time every shareholder has for irrigation, they have invented a special type of water clock or clepsydra. Their clepsydra consists of two bowls made of copper one of which is so small that could freely float on the surface of water in the large one. The floating bowl has a tiny hole at its bottom through which water can enter the bowl and gradually fill it up. After being filled which may take a certain time, the small bowl sinks in the water and bumps into the bottom of the large bowl. As soon as the bump would be heard, a unit of time would be over, so the time between the two bumps equals a certain unit of time. One can also find some marks cut into the inner side of the small bowl which divide the certain unit of time into the shorter fragments. The time it may take the small bowl to be filled and sink varies from area to area in the central plateau of Iran (Fig. 1.7, Table 1.1).

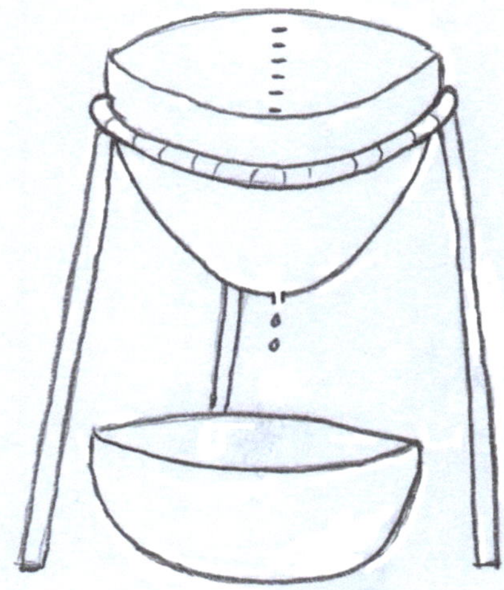

Fig. 1.7 Traditional water clock for calculating the irrigation time

Table 1.1 Duration of time unit in some parts of Iran

Location	Time (hour: minute: second)
Kol-e Birjand	00: 24: 00
Shahik-e Qayen	00: 22: 30
Khor-e Birjand	00: 17: 00
Kadekan	00: 15: 00
Sarbisheh, Zirkooh-e Qayen, Darmian-e Birjand	00: 12: 00
Yazd	00: 11: 15
Zoozan, Boshrooyeh	00: 10: 00
Fakhrabad-e Bajestan, Eshgh Abad-e Tabas	00: 09: 00
Bilond-e Gonabad	00: 08: 30
Gonabad	00: 08: 24
Dihook-e Tabas	00: 08: 00
Khanik-e Gonabad	00: 07: 30
Abiz-e Qayen	00: 07: 00
Aboojafari-e Boshrooye, Kakhk	00: 06: 00
Khosro Jerd-e Sabzevar	00: 05: 00
Serend-e Ferdows	00: 04: 44
Bajestan	00: 04: 36
Tabas	00: 04: 00
Ferdows	00: 03: 00

As mentioned, due to the complexity of water division, there are some professionals named "Mirab" who are in charge of distribution of water among the farms, and are paid a certain salary by all the shareholders. Mirab should be quite familiar with the network of ditches leading water to the farms, because he has to factor in the time that water takes to travel the ditches and reach each farm. Mirab has a notebook too, including all the irrigation rights in detail, so if the shareholders want to sell or buy any right they should inform mirab of any transaction. Unfortunately nowadays this profession is fading and nothing is replacing it, so we witness some recent conflicts over water in rural areas these years.

1.5 Qanat Civilization

In this book we do not want to delve deep into the concept of qanat civilization, but in a nutshell we can only say that Qanat civilization came into existence in such arid and semi arid regions as central Iran in the wake of climate change in the area, making the available surface water resources diminish. The shortage of surface water forced the ancient communities to turn to the groundwater as a substitute which could perpetuate life and prosperity in the region. Therefore qanat emerged as a response to the climate change and then surface water scarcity. Qanat was a man made means to extract groundwater, so it always needed maintenance to be able to keep operating, unlike a natural stream which runs without human intervention. Qanat discharge is limited altogether due to its technical specifications, and this water could only irrigate a local agriculture with low income in comparison to the vast agricultural systems based on permanent rivers. All these factors underlay a fragile economy which was much dependant on qanat water as well as handicraft to supplement the low income of agriculture. Nevertheless, the local agriculture based on qanat water and handicraft could not come up with a surplus production which was necessary to develop military and political structures. That is why one can rarely find the trace of extensive political and military systems and accordingly huge battles and conflicts in the central desert areas of Iran. In other words, the small discharge of qanat sufficed only to irrigate a limited area which did not allow much revenue. Lack of surplus production did not favor the development of huge military and political systems which required a considerable budget. On the other hand, the population of the arid lands preferred not to risk their qanats by waging war on their neighbors or provoking them to invade, since qanats were very vulnerable and could be totally blocked only by filling up one of its wells. The town's resistance could be broken easily when its qanat was destroyed or poisoned, since it was the only water source without which life was impossible. Therefore, qanat gave rise to a prudent society living off a fragile economy, known as qanat civilization which embodies connivance, leniency and tolerance (Papoli Yazdi and Labbaf Khaneiki 2000: 1–23) (Fig. 1.8).

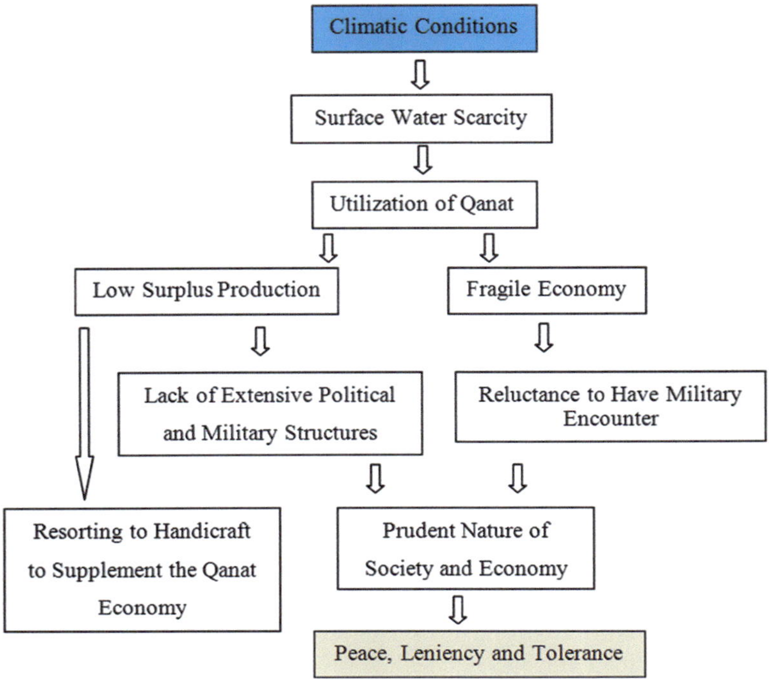

Fig. 1.8 Causal relationships forming qanat civilization

1.6 Qanat and Wittfogel's Theory

Government's intervention in water management in the history of arid and semi arid regions brings us to the conclusion that power and water are intertwined. But it seems difficult to explain the subtle relation between power and water. It was possible that scarcity of water made ancient people concentrate around the available water resources, and they had to invent a social management system to ration water, and later this system came to underlie political structures. This theory immediately reminds us of Wittfogel who coined the thought of hydraulic agriculture and civilization, and of some other scholars like Jared Diamond who considers such environmental factors as water the main determinants of the rise and fall of empires (Diamond Jared 2005). Even later Worster built on Wittfogel's theory to argue that the American West is not the center of individualism as it is made out to be, but is a hydraulic society, dependent for its current existence on the ability of a complex and expensive network of irrigation systems for its production, and thus dependent upon the capital and the authority by which those irrigation systems were built and are maintained. Worster used the Wittfogel's model once set out about the ancient Chinese or Indian empires to give insight into the socio-political mechanism of the American West. He went down the same path as that of Wittfogel to explain how taking control over water can bring about a powerful social class. According to him, "control of water meant wealth and power, while lack of control meant relegation to the lower classes, struggle for existence, and a lack of individual choice" (Worster 1985).

This logic is not only confined to Wittfogel or Worster, but dates back to hundreds of years ago, if we keep tracking it down in the books and records. The Iranian geographer Narshakhi (943 AD) writes in his book about the geneses of Bokhara[3] which was one of the most well known cities at his time, and he chalks up the creation of the city to the existence of water in that region. He believes that there existed a river that attracted wandering people and they got settled around the river and became civilized and later they had to assemble and live under a monarchy. He says: "There was a huge river running toward Samarghand, which was called Masef. A large amount of water flowed down that river and washed away much soil and sand which filled up all the holes and caves and paved the land over. Later this land was named Bokhara and the river which ran across it was named Soqd. People gathered around that river and prosperity came to that land over time. Many people came from Turkistan in search of water and easy life. First they put up some tents and lived like nomads for a while, but later they came to build homes. The population grew larger gradually until they singled out someone from among themselves to be their ruler named Abrovi" (Narshakhi Aboobakr, 943 AD: 7). According to Narshakhi, the river attracted the population and paved the way for human settlements that lived off water and irrigation. Irrigation on a large scale entailed a managerial authority that could oversee water division as well as maintain the ditches and irrigation networks. Thus the first ruler of Bokhara came into existence out of water.

On the other hand, there are some studies refuting what Wittfogel or others set out about the causal relation between water and political structure in the so-called hydraulic civilizations, and this controversy continues unabated. For example some archeological studies in Mesopotamia show that irrigation has only been related very indirectly to the advent of the state, nearly 5000 years later. Adams explains that in Mesopotamia large waterworks are developed long after the state (Adams 1965: 40), while Gibson argues that irrigation leads to nothing but economic disaster (Gibson 1974). According to some scholars, "the early development of irrigation canals was a local affair, carried out at a limited scale by local communities. Eventually the local networks were connected to wider systems, so that the growth of hydraulic structures went hand-in-hand with the growth of political systems and was not the result of centralized polities but rather a factor in their development" (Pollock and Bernbeck 2005: 245).

If we get back to Wittfogel's theory and his advocates, we have to focus on the difference between qanat as a small scale irrigation with bigger water bodies which necessitated a great deal of labor to transfer, distribute and then manage the water flow. The government's intervention took place mostly in terms of big water resources such as permanent rivers, and in case of small resources such as qanats the role of people and private property became more discernible. Even Wittfogel

[3] Bokhara (Buxoro) is the capital of the Bokhara Province of today Uzbekistan.

himself points out this difference where he writes: "hydroagriculture, farming based on small scale irrigation, increases the food supply, but it does not involve the patterns of organization and social control that characterizes hydraulic agriculture and oriental despotism. These patterns come into being when an experimenting community of farmers or protofarmers finds large sources of moisture in a dry but potentially fertile area" (Wittfogel 1957: 18). We can barely find a qanat system or even a group of qanats which could have underlain a national government. In the central plateau of Iran where qanats are the cornerstone of local economies, no national power has originated. The management of qanat is as accurate and intricate as that of the big rivers, but it has never led to what Wittfogel says about the relationship between water and power because:

1. In most cases a qanat was constructed by a number of farmers and later was owned by the same people. The water management was run by a group of people or by someone who was hired by the qanat shareholders.
2. A qanat could not drain so much water that a large area was possible to be irrigated and cultivated. Therefore the revenue of qanats was limited and did not leave a surplus in order to pay for a political or military system.

Therefore qanat civilization does not much tally with what Wittfogel attributes to hydraulic civilization though it is based on water too. Qanat irrigation did not give rise to a tyrant taking control of water and accordingly agricultural production systems, because qanat used to be run and managed by all the shareholders. Qanat management was premised on the sense of cooperation among the shareholders, and in this system everyone's voice could be heard. In most cases, the qanat's shareholders congregated in a place once a year in order to elect someone responsible for dividing water among them. They could easily oust that person if they were not content with his job. So nobody could find an opportunity to grow into a water king exerting authority over the others.

On the other hand, most qanats need to be repaired and cleaned out at least once a year to function properly. Cleaning out a qanat with such an extent required contribution from all the shareholders and a single individual did not afford to undertake this job alone. Therefore qanat civilization enjoyed some characteristics which signify a civilization in sharp contrast to what Wittfogel portrays in his work, a civilization which did not accommodate any feudalism or national kingdom in order to draw on the resources of the other nations. Qanat civilization was the civilization of peace, cooperation and friendly coexistence.

1.7 Qanat in Comparison with Other Water Related Structures

Altogether water is the most abundant element on the earth and is also the most serious problem that humans have ever faced in the course of their history, a problem that used to overwhelm humans because of whether its influx or its scarcity. We set foot in the world with plenty of water which makes up 80 % of our body as infant.

Somewhere between 70 and 75 % of the earth's surface is covered with water. Therefore we are surrounded all the way by water in our environment and are also awash with water in our own bodies. That is why water is an element whose footprint is invisible in our culture the most. Almost every culture used to have and worship a water deity whose duty was to protect water supplies for humans or to protect the humans against the devastation of water or both. Our human cultures and history have been inundated with water goddesses or gods with different names which imply the importance of water in human mind and culture, for example Chalchiuhtlicue in Aztec Mexico, Acionna in Celtic Culture, Mazu in China, Tefnut in Egypt, Vedenemo in Finland, Anapos in Greece, Apam Napat in India, Pariacaca in Incan culture, Suijin in Japan, Abzu in Mesopotamia, Volturnus in Rome and eventually Anahita in Persia. For the both purposes – protection of water for humans and protection of humans against water – the ancient people came to invent a variety of technologies each of which enjoys a long history of utilization and a deep impact on human culture. All the water related technologies and structures fall into 4 groups as follows:

1. Technologies for obtaining water
2. Technologies for transferring water
3. Technologies for storing water
4. Technologies for warding off water

Now we want to leaf through hundreds of pages written about the above mentioned technologies to better know why qanat still stands out as a technology that has been paving the way for many people to live in as harsh environment as the Iranian desert. In the group 1, qanat does not stand alone, but there are also other technologies which supply or supplied water to the human settlements. For example no one can deny the significance of historical dams which made it possible for people to make the most of the small streams. A dam should not be equated with a simple structure for storing water, but actually a dam enables people to rely on a fluctuating stream, otherwise they may be either drowned in plenty of water or die out of thirst. The technology of dam can help supply water in a more constant and reliable way rather than living through an "either famine or feast" policy. We can track down the oldest dams in the Middle East and North Africa where the first farming sedentary communities came into existence. Jawa Dam is regarded as the first gravity dam that dates back to 3000 years before Christ. This ancient dam is located 100 km northeast of the capital Amman in Jordan, and is 9 m high and 1 m wide, supported by a 50 m wide earth rampart (Garbrecht 1986: 51–64). The Egyptian Sadd-el-Kafara Dam at Wadi Al-Garawi is another example which dates back to around 2600 (Bazza 2006). Most of the dams were built across the big rivers which could irrigate a vast area of farmlands. These structures entailed a great deal of energy and money which did not be afforded by single individuals or small communities. Therefore the governments and actually the ancient monarchies were usually behind the construction of those dams, unlike the qanats that were always built and owned by petty landowners. That is why sort of decentralization of power and economy is inherent in the technology of qanat, which distinguishes it from such hydraulic structure as dam.

Another technology that fits the group one is the water harvesting wells which can be found in the southern parts of Iran. These wells are dug in rocks and lime stones in order to receive the seasonal runoffs whenever it rains. The wells filled with rain water can provide the people with potable water for the dry months when the sky becomes stingy about rain. The wells of Laft in Hormozgan province, Iran were dug in the Achaemenid period for this purpose. The rain harvesting wells have been serving to supply potable water, but this task is only one of the several tasks that a typical qanat can fulfill. Qanat is not only used for supplying potable water, but it is also used for sanitation, irrigation of farmlands, generating of energy, etc. The multitasking nature of qanat helped those people to better adapt themselves to their arid environment.

Qanat still holds its importance when it comes to the technologies for transferring water. One of the famous technologies of this kind is the ancient aqueduct which was built for spanning two sides of a barrier like a gully or a valley where the water current had to cross on its way. The best example of this technology is always attributed to Romans, but aqueducts were also used as early as the seventh century before Christ when the Assyrians built an 80 km long limestone aqueduct, which included a 10 m high conduit to cross a 300 m wide valley. This aqueduct once served to carry water to the city Nineveh (Jacobsen and Lloyd 1935). Aqueduct and qanat are similar in bringing water to the vicinity of humans where all the soil and climatic conditions as well as security – except for water -favor establishing the settlements rather than bringing the settlements to the vicinity of water. Their resemblance is perceptible in their profiles in which the aqueduct turns out to be like a negative picture of a qanat where the solid parts become hollow and vice versa (Fig. 1.9).

Fig. 1.9 Reverse resemblance of qanat and aqueducts in their profiles

Nevertheless they have a crucial difference which has something to do with their technical characteristics. Unlike an aqueduct a qanat is a dynamic system which cuts through the soil and advances into the saturated area underground and even develops some side branches over time in order to get more water or at least keep its discharge steady. An aqueduct is built at once as an integrated structure in order to convey water from a place to another, but a qanat is built over a long period in order to keep pace with the aquifer receding. In the first place a qanat may be dug with the length of a few hundred meters, but it is extended to keep in balance with the groundwater and it may end up in tens of kilometers in length after centuries.

Madi in Isfahan, Iran is another example for the water transferring technologies. Madi is an artificial tributary derived from the river of Zayandeh Rood. The network of such tributaries distributed water among the farmlands in the region as accurately as possible. The engineering and technical aspects of Madi are not as intricate as that of qanat, but its management system and water division is akin to that of qanat.

Water reservoirs are among the Technologies for storing water. Ab Anbar is a Persian name for traditional water reservoir which is generally filled up with the water of qanat. Ab Anbar is an underground structure, constructed to store freshwater for drinking. The reservoir was fed from a nearby shallow qanat. All of the water reservoirs had a storage tank whose dimensions depended on the amount of qanat discharge and the demand for water. Ab Anbar or water reservoir in Iran is mostly contingent on qanat systems and qanat plays a crucial role in their function. In fact Ab Anbar is a part of qanat to help carry out one of the functions of qanat, which is procuring potable water.

People do not always go about inventing technologies to extract or store water, but sometimes they have to come up with ideas on how to get rid of the extra water. One of the famous technologies in this respect is the windmills of Holland. Those windmills historically served many purposes. The most important probably was pumping water out of the lowlands and back into the rivers beyond the dikes so that the land could be farmed. The Dutch have become very innovative when it comes to keeping out the water. They have built dykes, fortifications and then windmills to create new land. The Dutch windmills carried water out of their lands and the qanats carried it to their lands both in order to facilitate farming and life. It is worth noting that unlike a windmill, a qanat with a length of kilometers cannot be operated and maintained by an individual or a few individuals. Cooperative utilization is in inherent in the technology of qanat, and is also manifest in its intricate water management and maintenance systems.

References

Adams Robert MC (1965) Land behind Baghdad, a history of settlement on the Diyala Plains. University of Chicago Press, Chicago

Bazza M (2006) Overview of the history of water resources and irrigation management in the near east region. Food and Agriculture Organization of the United Nations

Boucharlat R (2001) Iron age water-draining galleries and the Iranian Qanat. In: Proceedings of the first international conference on the archeology of the U.A.E
Briant P (2001) Irrigation et drainage dans l'Antiquité, qanâts et canalisations souterraines en Iran, en Égypte et en Grèce. Thotm Éditions (Persika 2), Paris, p 190, p. ill
Chauveau M (2001) Les qanates dans les ostraca de Manawir, Paris. Seminaire tenu au au colloge de France, Persika 2
Cheraghi Z (2010) Role of Qanats in the historic life of Yazd from Ilkhanid dynasty to Pahlavi I Era (1295–1941); a critical study. Isfahan University
Diamond J (2005) Collapse: how societies choose to fail or survive. Penguin Group, Camberwell
Garbrecht G (1986) Wasserspeicher (Talsperren) in der Antike, Antike Welt, 2nd special edition: Antiker Wasserbau
Gibson MG (1974) Violation of fallow and engineered disaster in Mesopotamian civilization. In: Theodore ED, McGuire G (eds) Irrigation's impact on society, vol 25, Anthropological Papers of the University of Arizona. University of Arizona Press, Tucson, pp 7–19
Goblot H (1979) Les Qanats, une technique d'acquisition de l'eau/English: Qanat a technique for obtaining water. Paris-La Haye, Mouton/Ecole des hautes en sciences sociales, p 236. Translated from French to Persian by Sarvqad Moqadam A, Papoli Yazdi MH (1992)
Hasanalian Davood (2006) Sialk; the unsolved puzzle of archaeology. Iran Newspaper. Archive at http://www.magiran.com/npview.asp?ID=1295401
Haupt L (1925) Armenien einst und yetst. Leipzig
Jacobsen T, Lloyd S (1935) Sennacherib's Aqueduct at Jerwan, Oriental Institute Publication 24. University of Chicago Press, Chicago
Kobori I (1964) Some considerations on the origin of the Qanat System, in memorial collected papers dedicated to Prof. E. Ishida, Tokyo
Labbaf Khaneiki M (2015) Qanat: a response to the climate change in Iranian Plateau. Asar Scientific Journal. Research Center of Iranian Cultural Heritage Organization, No. 66
Lightfoot DR (1996) Syrian Qanat Romani: history, ecology, abandonment. J Arid Environ 33(3):321
Lightfoot DR (1997) Qanats in the Levant: hydraulic technology at the periphery of early empires. Technol Cult 38(2):432
Narshakhi A (943 AD) edited by Hossein KA. Electronic Publication of Iranian History
Papoli Yazdi MH, Labbaf Khaneiki M (2000) The role of Qanat in formation of civilizations; sustainability of Qanat civilization and culture. In: Proceedings of the international conference on Qanat, vol 1.Yazd Regional Water Authority, Yazd
Pollock S, Bernbeck R (2005) Archaeologies of the Middle East critical perspectives. Blackwell Publishing Ltd, Malden
Salvini M (2001) Pas de Qanat en Urartu, Paris, Irrigation et drainage dans l'antiquite, qanats et canalisation souteraines en Iran, Egypt et en Grece
Semsar Yazdi AA, Labbaf Khaneik M (2010) Veins of desert. International Center on Qanats and Historic Hydraulic Structures (ICQHS), Yazd
Weisgerber G (2003) The impact of the dynamics of Qanats and Aflaj on oases in Oman: comparisons with Iran and Bahrain. Internationales frontinus-symposium, Luxemburg
Wittfogel KA (1957) Oriental despotism: a comparative study of total power. Yale University Press, London
Worster D (1985) Rivers of empire: water, aridity, and the growth of the American west, 1st edn. Pantheon Books, New York

Chapter 2
Hydrology Mechanism of Qanat

Abstract This chapter provides a technical definition of qanat and then examines the hydrology of qanat. Groundwater exists in different forms according to which qanat appears in different types. In this chapter we try to come up with a new classification covering almost all possible types of qanat according to different criteria. This chapter also goes over the general mechanism of qanat, and definitions of shaft wells, mother well, water production section, water transport section and qanat exit.

Keywords Qanat hydrology • Groundwater • Hydraulic gradient • Sediment • Seepage

2.1 Definition of Qanat

To simplify the issue, we had better define qanat as an almost horizontal tunnel with some shaft wells, which conveys groundwater to the earth surface. This water seeps into the tunnel in the saturated area and then flows down the tunnel toward the earth surface. In other words one can consider qanat a drainage system which drains out water in order to supply potable water and to irrigate the cultivated lands located down slope. As depicted in the below figure each qanats consists of two different parts; water production section and water transport section. Part of tunnel dug through the underground saturated area (aquifer) is called water production section and part of tunnel which just serves to convey water to the earth surface is water transport section. The border between these two sections is not stationary all the time and it moves to and fro with amount of precipitation and accordingly the expansion of aquifer. To have a better handle on the mechanism of qanat, we brief you on the concept of groundwater (Fig. 2.1).

A.A. Semsar Yazdi, M. Labbaf Khaneiki, *Qanat Knowledge*,
DOI 10.1007/978-94-024-0957-4_2

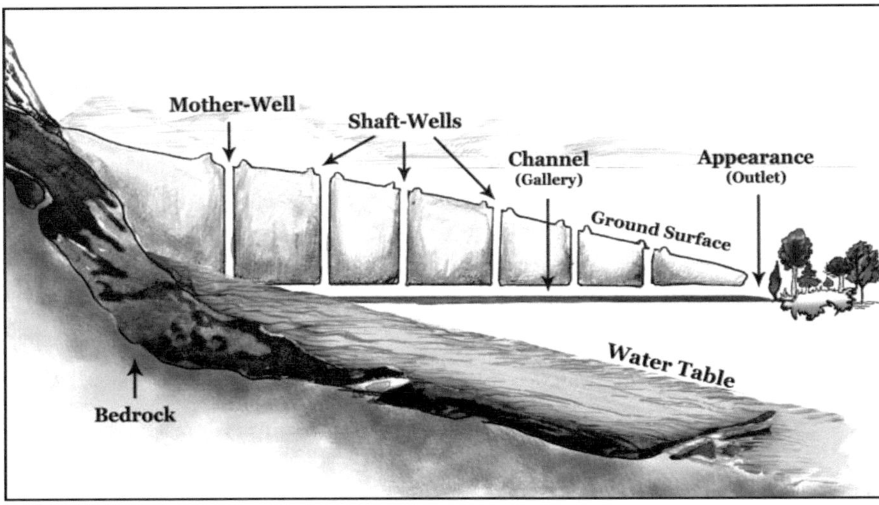

Fig. 2.1 Profile of a typical qanat

2.2 General Mechanism of Qanat

As mentioned, a typical qanat consists of a series of vertical shafts in sloping ground, interconnected at the bottom by a tunnel with a gradient flatter than that of the ground. The first shaft (mother well) is sunk, usually into an alluvial fan, to a level below the groundwater table. Shafts are sunk at intervals of 20–200 m in a line between the groundwater recharge zone and the irrigated land.

The qanat gallery is called in Persian "rahrow" or "kooreh", which is an almost horizontal tunnel dug to get access to groundwater reserves, and to transfer this water to the earth's surface. The dimensions of the gallery are such that the workers can easily go through and work in it: between 90 and 150 cm high and its width is less than half the height.

Taking into account the topography and position of the water table, the tunnel gradient is calculated so that the water flowing in the tunnel would have an adequate speed not to leave too much sediment and, on the other hand, would not be so rapid that it washes away the tunnel and causes erosion. The optimum gradient is taken to be between 2/1000 and 5/1000. The length of the tunnel depends on a variety of factors such as the topography, aquifer behavior, qanat discharge, where the water is to be used, etc. A qanat may be a few hundred meters to about 100 km long. The greater the extent of qanat in the water -bearing zone, the more water infiltrates into the tunnel. This is why a qanat may have several side branches in the water -bearing zone to increase the discharge.

The part of the qanat tunnel which cuts through the aquifer and collects the groundwater seepage is called "tare-Kar" or water production section. This part is below the water table and is fed by the aquifer. The discharge of the qanat completely depends on the extent, permeability and the depth of this part in comparison to water table. The length of the “tare-Kar” is not fixed but correlates with the fluctuations in water table.

The part of the tunnel which simply conveys the collected water to the surface is called “khoshke-kar” or water transport section. The border between the water production and transport sections is called “hadd-e taran va khoshkan” or transition zone. Where the tunnel and the ground surface eventually intersect is the exit of qanat which is called the “mazhar” meaning where water appears.

There are some vertical shafts sunk along the tunnel to connect the surface to the horizontal gallery. The main application of these wells is to lift out the spoil and excavated materials from the tunnel on to the surface. They also provide access and help ventilate the tunnel and provide more oxygen for the workers. These wells play an important role in repairing the qanat, by making it possible to send down the needed facilities and tools and remove the debris. The shaft wells cut short the time needed for qanat construction or repair, and reduce the relevant expenses. A shaft well is between 80 and 100 cm in diameter, and the distance between the wells vary from 20 to 200 m. The furthest shaft well from the exit, sunk upstream is called the “mother well”. The mother well is usually the deepest well, in which a large inflow of water shows the qanat is in a satisfactory state. If the water table goes down so much that it is located below the bottom of mother well, no water can seep into the gallery, and if this situation persists, the qanat will inevitably dry up. If a qanat is extended so far that another well is needed, the new well would now be the mother well and the former one would be a normal shaft well. In a nutshell, the last well is always called the mother well. The depth of the mother well varies from qanat to qanat, and the deepest one in Iran has been recorded in the qanat of Qasabeh Gonabad at 300 m.

2.3 Groundwater and Qanat

Groundwater is the water that exists in the earth fractures, faults or porous sediments. This water may origin from direct percolation of rainwater, distillation of water vapor underground, and eventually fossil groundwater. What extracted as groundwater is the water which comes from precipitation and percolates the earth and then forms the large aquifers. Such aquifers underlie qanat systems and a typical qanat functions as an escape way for the water trapped in the alluvial formations.

Aquifers can be classified into the following types according to their hydraulic behavior and physical traits:

1. Unconfined aquifers: these aquifers are limited to an impermeable layer lying beneath them. Their upper limit is considered water table and the pressure at that

point is equal to atmospheric pressure. In case a well would be sunk in such aquifers, the water at well and in the aquifer would be at the same level.

2. Confined aquifer: these aquifers are confined by two impermeable layers lying above and beneath the saturated area. Water in such aquifer is under pressure and would come up to regain its balance with aquifer if a well would be dug into the aquifer. These aquifers are unconfined only in the spot where they can be replenished.
3. Suspended aquifer: sometimes an impermeable layer in the shape of a lens is embedded in among the other alluvial deposits. Water can accumulate on this limited layer and form a suspended aquifer.
4. Semi-confined aquifer: where the saturated layer is encompassed by an impermeable layer from under and by a semi permeable layer from above. When water is under pressure it can seep into the upper semi permeable layer and then seep back into the aquifer when water is extracted and the pressure release. Therefore such aquifers are considered reliable.

Aquifers are dynamic water reserves which are replenished by precipitation and then discharged by different means annually. Qanat is a way to drain out groundwater without throwing the groundwater off balance. The water table lowers in the vicinity of qanat tunnel where it cuts through aquifer, and the level difference between these two points in the aquifer causes water to infiltrate into the tunnel.

Water cycle takes place according to simple principles and at the same shows intricacy. Since our childhood we have learned that sun shines over water bodies and water vapor ascends into the sky and forms masses of water vapor as clouds. When these clouds get cold and reach the dew point, they condense and then it starts raining. Precipitation takes place in the shapes of rain, snow or hail, some of which form the snow storage and glaciers in the mountains and some flow down the gullies, tributaries and ditches. Also a part of precipitation seeps into the earth and forms aquifers and another part evaporates and gets back into the atmosphere. The

Fig. 2.2 Position of qanat in water cycle

water that flows as rivers and runoffs ends up in the seas, lakes and oceans and the water cycle continues (Fig. 2.2).

Uneven distribution of precipitation all over the world has resulted in water scarcity in some lands with a desert landscape and in water abundance in some other lands covered with lush vegetation. In the arid and semi arid lands, humans have had to subsist on groundwater to ensure their survival in the harsh environments otherwise it has been impossible to live on in the hope of surface water. In the moderate and rainy regions humans have relied on the permanent surface waters to live on. The remains left from the ancient civilizations prove that humans have strived to harness and exploit these resources by inventing a variety of technologies. The question is that where in this water cycle the qanat system stands? In other words what role the qanats play in the water cycle in nature? It is not that difficult to answer this question; most qanats consist of a subterranean tunnel which drains out groundwater and transfers it to the earth surface. This water can be either the water stored in the rich deep sediments or the water brought by a fault or underground stream or the water seeping from a shallow aquifer. Moreover there exist other kinds of qanat such as river-qanat, spring-qanat and the qanats that condense the water vapor of the air. In general one can mention two different types of qanat according to their positions in water cycle as follows:

Type 1 – Infiltration qanats: these qanats in fact consist in an artificial groundwater mining system which is built into the water saturated sediments to extract its water and convey it onto the earth surface. In other words these qanats play a role in the water cycle by bringing groundwater to the earth surface through their tunnels whose gentle slope allows them to intersect with the earth surface. These qanats can be classified into the following groups according to kind of water that they extract.

(A) Extracting the integrated aquifers: many of qanats in Iran and other countries are among these qanats which have been dug in the plains with extensive groundwater reserves. These qanats cut through the rich aquifers and drain out groundwater toward the earth surface. Given that surface slope is gentler in such regions and generally the aquifers are deeper, these qanats are deeper and accordingly longer. These qanats enjoy a relatively constant flow with minimum fluctuation in their discharge, due to their access to constant deep aquifers. In fact these qanats are fed by vast and deep aquifers with considerable water storage, so any fluctuations in the amount of precipitation cannot affect them in short term. That is why these qanats provide a constant discharge during the different seasons of the year. The following picture shows the profile of a typical qanat of this kind (Fig. 2.3).

These qanats enjoy special structural and technical characteristics, due to their high discharge, depth, extent and considerable length which passes through different earth features and geological formations. In other words, the technologies used in such qanats are more intricate than that of other qanats, the technologies which depict the human ingenuity. One of these technologies is digging a qanat tunnel and its wells below the water table

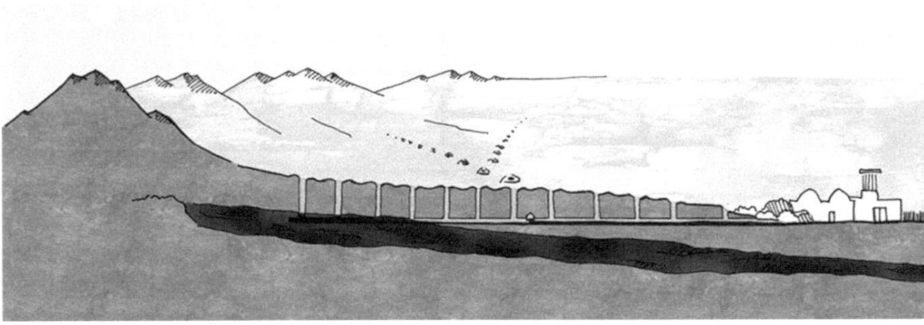

Fig. 2.3 Profile of a qanat dug into a plain extensive aquifer

which are called "Devil" in qanat workers' jargon. This method has been described in detail in the next chapters. In comparison to the other types of qanat, these qanats enjoy more sustainable and intricate water management, exploitation and maintenance systems. In fact such systems paved the way for these qanats to endure over the past centuries.

(B) Extracting groundwater streams: There are some underground ravines and faults which convey water from the mountains to the less elevated lands and plains. These ravines serve to recharge the aquifers. In case a qanat tunnel happens to intersect one of these underground ravines, their water would be diverted to the tunnel. Such qanats often enjoy a high discharge which is of high quality, since they extract a free flow of groundwater. In Iran in Bam there exist a large number of such qanats some of which drain out 250 l per second of groundwater. They are ranked among the qanat largest discharges. In the picture below you can see a typical qanat which is fed by the underground streams (Fig. 2.4).

(C) Extracting shallow aquifers: the mountainous areas which are considered source of recharge for the plains downstream usually enjoy shallow and limited aquifers which closely correlate with the amount of precipitation. In fact the alluvium lying on the mountain base is a passage way for the water to flow down toward the plain. After a precipitation the water flows down the elevated lands toward the low lands according to the surface slope and forms groundwater reserves there. Therefore in the mountainous areas the bed rock does not lie deep and accordingly there is not a deep layer of sediment to form a rich aquifer. That is why such aquifers are fully dependant on the amount of precipitation. Now if a qanat is built in such aquifers, this qanat is regarded as a mountainous qanat which is relatively short in length,[1] and the shallowness of sediment makes its discharge fluctuate. In other words if precipitation would be high the qanat discharge would increase immediately, and if a drought would pervade the region the qanat discharge would sharply drops or even dries up. Such qanats are locally called "Hava

[1] Because of the steep slope of the gound, the qanat tunnel and ground surface intersect soon.

Fig. 2.4 A typical qanat which extracts the underground streams

Fig. 2.5 A typical "Hava Bin" qanat in mountainous areas

Bin",[2] which are very vulnerable to any change in the amount of precipitation. Their discharge reaches its peak after each rainfall and drops to its minimum at the end of summer. Therefore these qanats have fluctuations throughout a year. In general these qanats do not enjoy a considerable discharge. The following figure portrays a qanat of this type (Fig. 2.5).

Though these qanats are fluctuating over a year and they may even dry up for a while, they come alive again whenever precipitation is enough. But the plain qanats are more stable but harder to become active again after drying up. The plain qanats are fed by thick and deep aquifers which have formed over the past hundreds of years, so over-exploitation of such aquifers leads to their depletion which cannot be fixed that soon. If the water table drops below the qanat gallery, the qanat is doomed to annihilation. That is why most of the abandoned qanats are plian qanats, whereas many mountainous qanats could survive and remain active. Nevertheless if the global warming and climate change persist in the coming years, more severe droughts would

[2] "hava-Bin" literally means "looking at sky" implying that such qanat discharge fully correlates with the amount of rainfall.

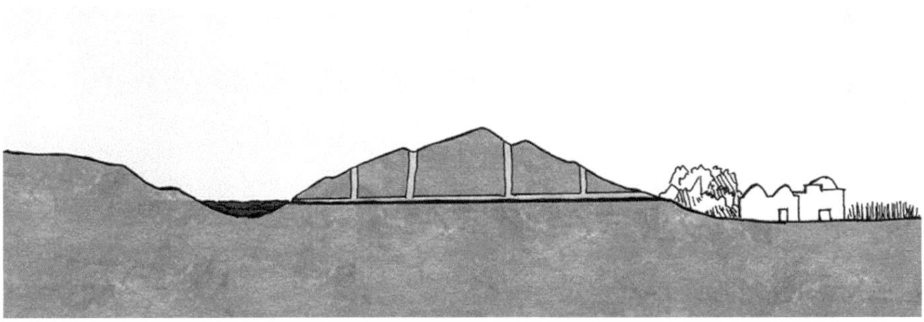

Fig. 2.6 Qanat type 2 which transfers water with no groundwater seepage in its gallery

pervade these regions and lead to a dramatic decline in the discharge of mountainous qanats.

Type 2 – Conveying qanats: these qanats just consist of a tunnel which branches from a river to transfer its water from beneath an elevated land lying between the river and the region to be irrigated. Therefore these qanats do not gather groundwater seepage, but just convey water. This type is built not only in the arid regions but also in such rainy areas as European countries. This type is a perfect method to convey water where a hill would lie between the water source and the desired land. As Abdullah Al-Ghafri elaborates in the book "Qanat in its Cradle", in Oman such qanat is called "aini" that draws water from one or more natural springs. The water flows from springs into the qanat channel, which transports it to agricultural lands. For example falaj Ain al-Kasfah in al-Rustaq in the al-Al-Batinah Region, and falaj Bowsher in the capital area transport water from the spring to the village for domestic and agricultural use. The aini aflaj represent 28 % of the total number of qanats in Oman (Semsar Yazdi and Labbaf Khaneiki 2011: 201) (Fig. 2.6).

2.4 Classification of Qanats

There are four criteria according to which one can classify qanats; 1. length and depth of qanat 2. topographical condition 3. geographical situation 4. aquifer type and qanat discharge 5. discharge origin

1. Length and depth of qanat: length and depth of qanat is the most common criteria which is in fact contingent on the region topography. According to this criterion qanats are classified into two types; long and short qanats. Short qanats run down the mountain foots and shortly meet the earth surface, whereas long qanats are located in the flat plain where the qanats have to be longer to meet the earth surface. The other criterion is qanat depth according to which qanats are classified into two groups of deep and shallow qanats. This criterion is associated with the depth of the impermeable layer on which water can build up after seeping into

the ground, so the qanat depth correlates with how deep the impermeable layer lies.

2. Topographical condition: Qanats are distributed across a variety of topographical conditions from mountainous areas to plains. According to this criterion we have three types of qanats; mountainous, semi mountainous and plain qanats. For example in Iran in Yazd area the earth becomes more elevated from the sea level as we approach Shirkooh mountains whose highest point is 4055 m. The qanats in Yazd surrounding are among the plains qanats, but we come across semi mountainous and mountainous qanats on our way to Shirkooh mountains.

   2.1 mountainous qanats: these qanats enjoy a relatively shallow mother well and accordingly shallow shaft wells. These qanats are fed by shallow aquifer. Steep slope of ground surface and shallowness of shaft wells cause the mountainous qanats to be short, and the gallery intersects the earth surface after a short distance. Therefore the workers have to dig a steeper tunnel than normal in order to allow the gallery to be longer and go further into the aquifer. The further the gallery goes into the aquifer the more water seeps into the gallery. Nevertheless the mountainous qanats cannot have as steady discharge as the plain qanats do. Shallow wells, short galley dug into aquifer, shallow bedrock and accordingly shallow aquifer lead these qanats to have a fluctuating discharge which closely correlates with annual precipitation.
   2.2 Semi mountainous qanats (mountain foot qanats): these qanats have a mother well dug in mountainous areas but their exit appears in the low lands and plains. These qanats are longer than the previous type so they enjoy more discharge though it is still fluctuating.
   2.3 plain qanats: these qanats are longer than the previous ones and enjoy a deep mother well. Given that in the plains the bed rock lies deeper, aquifer forms at deeper layers. Deep wells and gentle surface slope cause the qanat gallery to be relatively long, which leads to a more reliable discharge.

3. Geographical situation: this concept refers to the geographical locations of some qanats in relation to each other. According to this criterion, we can recognize three different types of qanats as follows:

   3.1 Successive qanats: such qanats are found mostly in mountainous areas especially along the steep valleys, so that each qanat can feed the next qanat down slope. In other words each qanat enjoys a mother well which has been sunk in the farmlands watered by another qanat up slope, and can get its water percolation. In the dry valleys with poor aquifers the discharge of qanats dwindles as we go down slope, because these qanats depend mostly on the water coming from irrigation. But in the valleys with better aquifers the down slope qanats have more discharge because they take advantage of irrigation water seepage in addition to the groundwater directly seeping into their galleries.

3.2 Parallel qanats: these qanats usually run down the mountain foots parallel to each other. These qanats originate from mountain foots overlooking a vast plain with thick layers of sediments. Precipitations on the mountains recharge the aquifer and seep into the qanats. In this case villages nestle on the mountain foot in a row and utilize the parallel qanats.
3.3 Converging qanats: these qanats exist in the lands which are surrounded by mountain ranges. In this case the qanats originate from the surrounding elevations and run toward the plain center, because groundwater flow is in fact radial, originating from the surrounding elevations toward the plain center.

4. Aquifer type and qanat discharge: Another criterion to classify qanats is qanat discharge according to which qanats are divided into two types; qanats with permanent discharge and qanats with fluctuating discharge. This criterion is associated with the type and size of aquifer to which a particular qanat is connected. Even one can classify qanats on whether a particular qanat cuts through a confined, semi confined or unconfined aquifer.
5. Water origin: it may seem strange to classify qanats according to the origin of water coming out of their tunnel, because it is taken for granted that water comes from saturated soil layers. But in fact some qanats convey the water that has nothing to do with groundwater direct seepage into their gallery, but this water may come from such sources as tube wells, natural springs, surface streams, or even other qanats. So this concept hinges on the fact that whether a particular qanat depends on groundwater infiltration into its own gallery or other sources. Thus one can classify qanats into the four types according to the aforementioned criterion: simple qanat, qanat-spring, qanat-river, qanat-well.

5.1 Simple qanat: simple qanat drains out only the water which seeps into its own gallery. The typical qanats are of this kind and the standard definition of qanat refers to simple qanat.
5.2 Qanat-spring: these qanats receive spring water through a special well locally called "Zorna". For example qanat of Ahrestan in Taft (Yazd, Iran) conveys half of the water of Tamer spring along with its own water.
5.3 Qanat-river: in this case a surface stream is directed into a qanat nearby through a well which has been constructed for this purpose (Zorna). To prevent erosion, river water is led to a pool so water settles first and then they let the water enter the qanat through Zorna. Some qanats are totally dependent on surface streams and no groundwater infiltrates into them. In Dezfool, Khuzestan province qanat system is employed to transfer river water to cultivated lands across some elevated areas in between. These qanats are called "Qomesh" in the region.

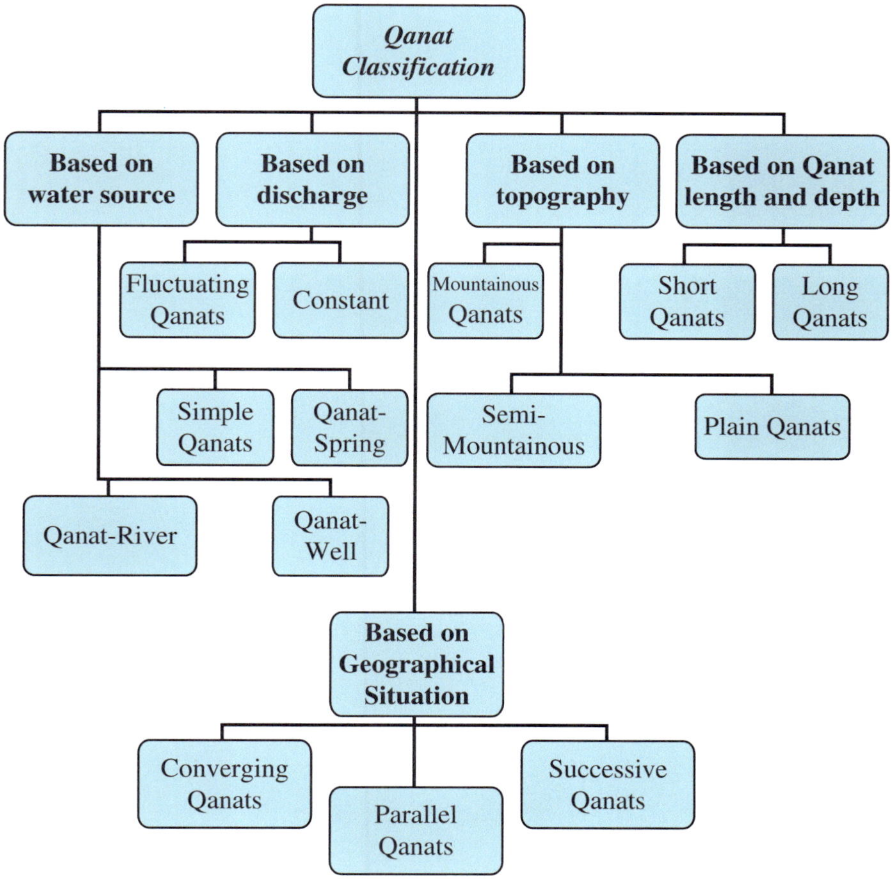

Fig. 2.7 Qanat classification

5.4 Qanat-well: sometimes water pumped by a tube well enters a qanat and mixes up with the qanat water. In some villages the owners of a tube well use qanat tunnel to transfer their water to their farmlands downstream. In this case they measure their water before entering the qanat and then receive the same amount at the qanat exit with the consensus of qanat owners (Fig. 2.7).

Reference

Semsar Yazdi AA, Labbaf Khaneiki M (2011) Qanat in its cradle, vol 1. International Center on Qanats and Historic Hydraulic Structures (ICQHS), Yazd

Chapter 3
Locating Qanat

Abtract This chapter examines the first step one should take to get started on qanat construction. We try to explain how to find a suitable place to dig a qanat based on traditional know how. There are a variety of indigenous methods to do so, some of which may be considered baseless from scientific point of view, but they still retain their ethnological values. In this chapter we show how the qanat masters have to factor in groundwater quality, groundwater depth, location of test wells, etc. while determining where to dig the qanat. Locating a qanat is a process which is described in this chapter. When a qanat master decides to dig a new qanat, he has to take into account such different elements as geological situation and natural indications.

Keywords Desert vegetation • Geographical stratifications • Aquifer condition • Test well • Water seeking

3.1 Natural Indications

The first thing that a qanat master considers to locate a qanat is the ground gradient which dictates the depth and length of qanat. In a steep land, the qanat gallery and ground surface intersect soon, but in a land with gentle slope the gallery has to travel a long distance to meet the surface. Therefore the master calculates the surface slope and then predicts whether the qanat water is to appear in the right place.

Qanat practitioners use the natural indications found in the area to locate a new qanat. For example gullies, seasonal rivers, mountain foots, alluvial fans, etc. are among the signs which imply that a particular area enjoys a reliable groundwater reserve. Also lush vegetation gives a clue to the existence of groundwater at a reasonable depth. The qanat masters are knowledgeable of the characteristics and life cycle of the wild plants growing in their vicinity. For example if they see a concentration of Alhagi sp.[1] somewhere in the desert, they conclude that there is groundwater

[1] Alhagi (camel thorn) is an invasive, perennial shrub in the Legume family with fairly deep creeping root. It is used in Iranian herbal medicine and is known for its antiasthmatic, aphrodiasiac, antipyretic, diuretic, expectorant and laxative effects. Its flowers are used to treat piles, migraine, and warts (Honari Maryam et al 2014: 268).

A.A. Semsar Yazdi, M. Labbaf Khaneiki, *Qanat Knowledge*,
DOI 10.1007/978-94-024-0957-4_3

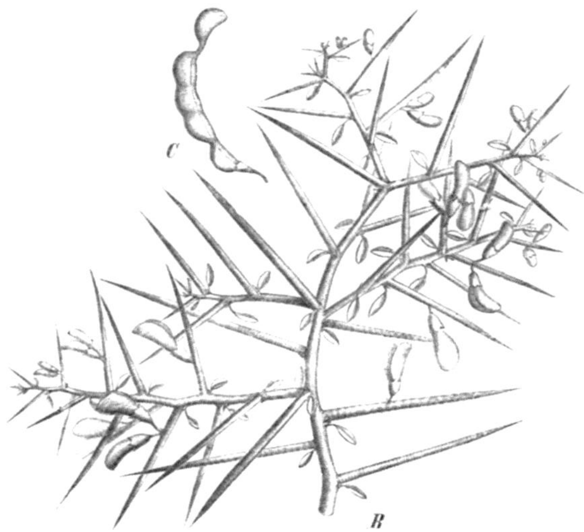

Fig. 3.1 Alhagi, a plant commonly used to locate groundwater (Taubert 1981)

at a depth of less than 15 m, because the roots of Alhagi feed on fresh water at such a depth or less (Fig. 3.1).

Situation of soil, color of geological formations and order of soil layers help the qanat masters designate a suitable place for digging a new qanat. For example the soils with small particles have a higher capacity to keep groundwater. The qanat masters also use geological stratification to spot the impermeable layer, its conductivity, thickness and depth. They pay attention to a cross section of ground stratification at a ravine or cliff. If impermeable layer lies too deep, the qanat can be deep and accordingly the tunnel may run a very long distance to the surface. A new qanat rarely exceeds a few hundred meters, because it does not make economic sense to dig a very long tunnel taking much time, energy and money. Therefore the qanat masters try to find a situation in which their qanat reaches groundwater at a reasonable depth and length in order to bring the water to the farmlands as soon as possible, though this qanat may be extended year by year as the aquifer recedes. Needless to say the extremely long qanat like qanat of Zarch with a length of over 70 km have not been excavated over night, but over hundreds of years while the qanat has been operating.

If the farmlands already exist, the qanat masters should be careful not to dig the qanat exit down slope from the farmlands. On the other hand the masters are not willing to dig a qanat where the impermeable layer lies so close to the earth surface, because it forms a shallow aquifer which cannot store enough groundwater. Another issue which is at fore while locating a qanat is its vicinity. The qanat masters should observe the bound of the neighboring qanats. If the new qanat is to be built in a soft soil, it should lie at least 1500 m away from the qanats nearby, and in a hard soil this distance is 1000 m (Semsar Yazdi 2010: 312).

In an ancient book written by Karaji,[2] left from tenth century, there are many ideas very similar to what the qanat masters say even today. In his book he says: "the

[2] Abu Bakr ibn Muhammad ibn al Hassan (Husayn) Karaji (953–1029 AD) was a Persian tenth century mathematician and engineer. He is considered one of the founders of Algebra in the Islam world (Rashed, 1970–80).

black mountains whose stones are mixed with mud are very likely to contain groundwater. After the black mountains, the green, yellow and red ones respectively enjoy groundwater most. The black mountains with soft layered rocks on their peaks have the most potential to contain water. The higher the ratio of stone to soil, the less groundwater is found in the mountain. The lone small hills have no water, especially where they are encrusted with hard rocks, because snow cannot last long there and percolate the soil. The mountain ranges with many valleys and ravines can tuck away the snow until spring and summer, so they have much water whatever color they are, especially if they enjoy flat peaks and lush vegetation on their slopes, because the plants can protect the water in soil against evaporation. In such mountains, the northern slopes are more likely to have water … whoever does not know the indications of groundwater on the surface cannot be a qualified muqqani (qanat master). The objects on the ground surface bear witness to the existence of water underground, and it suffices for the people to look carefully. All the plains connected to the aforesaid mountains are awash with groundwater, especially if their soils are porous. The farther from the mountain toward the center of the plain a place would lie, the shallower the groundwater there. If a place is covered by lush vegetation there is a good reserve of groundwater there, especially where the leaves are damp from the morning dew. In the grounds with gullies cut by the runoffs that flow down the nearby mountains, one can find groundwater especially if the runoffs have no way but to seep into the ground. In case the ground surface is damp from the morning dew, that place is expected to have groundwater. If wind howls in a ravine, one can find groundwater some distance down slope provided there would be traces of plants and dew there, otherwise it is just wind blowing … Some plants heralded the existence of groundwater, such as Portulaca, Solanum nigrum, Mentha pulegium, Rumex acetosella, Borago officinalis, Cirsium vulgare (Spear Thistle), Urtica dioica, Glycyrrhiza glabra, Alhagi, etc. The green robust plants hint that there is groundwater in there, unless those plants are watered artificially … I was told that someone dug a well in a place where Alhagi grew. The roots of Alhagi kept going down as the well was dug deeper until the roots reached the water at the depth of 50 Zeraa.[3] One of the best watermelons is one grown inside the roots of Alhagi. They make a cut in the root and put a watermelon seed in it so that the seed can feed on the water secreting from the wound until it grows leaves and flowers and then bears fruits. Nevertheless I do not believe that Portulaca, Mentha pulegium (squaw mint) and reed can lead us to groundwater, because their roots are not deep, and they live off the current waters or shallow stagnant waters" (Karaji 1966: 21–23).

The qanat masters may study the situation of the other groundwater resources nearby to deduce how their new qanat is going to be. If the qanats and springs in the area would be drizzling or fluctuating, they would not fasten their hope on a new qanat nearby as all are draining the same aquifer. In different parts of Iran there are some other methods as well in order to locate a new qanat. Some of these methods are conjectural or one cannot find a satisfactory scientific reason to give credence to them at least by now. Here we mention some of these methods which at least carry

[3] The Zeraa is an ancient unit of length based on the length of the forearm from the elbow to the tip of the middle finger, which equals 45 cm.

ethnological value. For example in some parts of Iran, in winter when the desert is encrusted by snow the qanat masters go out and look around to see where in the desert the snow would melt sooner, leaving muddy patches. They believe that groundwater is shallower at these patches because water vapor can reach the surface and melt the snow there. Sometimes they use a wad of wool to make sure if a good groundwater reserve exists underground. They dig a small hole and put the wool there and then place a bowl on the hole upside down to keep the water vapor inside. They leave the wool for a day and then they weigh it immediately to see how different it is from yesterday. The heavier the wool would become, the more water there would be underground. They use an accurate gold scale for this purpose. According to them, the groundwater vapor rises through the porous soil and then builds up under the bowl. This vapor is distilled after it is cooled by the air overnight, and the water drops soak the wad of wool and then make it heavier (Fig. 3.2).

Some qanat masters believe that wherever whirlwinds occur more often, it is more likely to find groundwater there. At the end of summer they go out to spot whirlwinds and they keep in mind where the whirlwinds take place more frequently. Even some of qanat masters are considered expert to use kind of "divining rod" to find water underground.[4] They whittle a Y-shaped rod off a nut tree and hold it loosely in their hands while walking gently. They believe that the presence of groundwater can affect the rod by making a feeble impulse which can be sensed under one's fingers. In this research we tried to find someone who is knowledgeable

Fig. 3.2 Wad of wool under an upside down bowl to predict groundwater

[4]The use of a forked twig, or so-called divining rod, in locating minerals, finding hidden treasure, or detecting criminals is a practice that has been a subject of Discussion since the middle of the sixteenth century and still has a strong hold on the popular mind (Ellis, 1917: 5). The Beausoleils are believed to have been influential in bringing about the use of forked twigs in searching for water, although Barrett writes as follows in regard to an account which he finds in a Life of Saint Teresa of Spain: Teresa in 1568 was offered the site for a convent to which there was only one objection there was no water supply; happily, a Friar Antonio came up with a twig in his hand, stopped at a certain spot, and appeared to be making the sign of the cross; hut Teresa says, "Really I cannot be sure if it were the sign he made, at any rate he made some movement with the twig and then he said, 'Dig just here'; they dug, and lo! A plentiful fount of water gushed forth, excellent for drinking, copious for crashing, and it never ran dry" (Ellis, 1917: 15).

Fig. 3.3 A qanat master using divining rod to locate groundwater

of this method in detail but we failed. It seems using divining rod has long been abolished at least in the region of Yazd. Michael Bonine mentions using this method in his research conducted between 1970 and 1977, though he does not seem to witness this method in person and probably he has come across this story during his interviews with locals. "In the past, some ostad-i muqanni [qanat masters] have used a divining rod made from a hazel-nut tree to help locate a source of water" (Bonine Micheal 1982: 146) (Fig. 3.3).

3.2 Digging Test Well

A qanat master does not go about building a qanat without digging test wells, even though he is sure about the existence of groundwater. Qanat construction is very costly and time consuming, and the qanat master does not get started until he makes sure about the quality and quantity of groundwater up close by digging test wells.

The qanat master figures out where to dig a test well, taking into account the soil type and the gradient of soil layers. A test well is named "Gamaneh" which is sunk first of all in order to confirm the existence of groundwater and to assess its depth. In other word after finding a suitable place for building a qanat through natural indications, the first step is to sink a test well. In fact there are many indications which can help the qanat masters locate an acceptable groundwater reserve and accordingly a suitable place for qanat construction, but it is still not fully certain to evaluate the hydrologic situation underground unless a test well would be dug. The qanat master can take it for granted whether or not a particular place is suitable for building a qanat after investigating the depth of test well, its infiltration rate as well as the surface slope. In many cases more than one test well are dug in order to get a better view on the situation of groundwater and make sure about the feasibility of qanat construction. Sometimes a test well may reach a limited groundwater current which does not represent the situation of the whole aquifer. In this case the qanat masters manage to dig three test wells in a row, some 500 or 700 m away from each other.

According to Mohammad Reza Fayyaz, a qanat master from Yazd, they dig a test well some one meter below the water table to make sure about the quality and quantity of groundwater. Afterward two other test wells are dug on the both sides of the first one to form a triangle, each 300 m away from the first test well. The forth test well is dug some distance up slope from the first one, in the same direction that the qanat is to be. Now they monitor the water level in the four wells to find out about the hydraulic gradient and the general current of groundwater (Semsar Yazdi 2010: 222).

In fact they measure the depth of these wells and the amount of water in them to find out about the extent and capacity of the aquifer. According to the qanat masters of Taft, a good test well should hit groundwater at a depth of between 30 and 35 m. If a test well reaches a shallow groundwater, such a qanat would be going to have an unreliable fluctuating discharge, locally called "Havabin". Of course the optimum depth of test well is associated with the geological and topographical conditions of the region. For example in Yazd in the area of Banadak, the village of Hassan Abad Akbari we found an undone test well left abandoned. The locals told us that they intended to build a new qanat, and they sank a test well which reached groundwater at the depth of 7 m contrary to their expectation. In fact there was an impermeable layer of clay at this depth, which held water in the shape of a suspended aquifer. If they trusted this water and continued building the qanat, the result was nothing but waste of time and money as the water dwindled soon and dried up eventually. Therefore they left the test well and the project fell through.

According to Haji Abbas Nasiri, one of the famous qanat masters in Yazd, the worker keeps digging the well even after reaching groundwater, and he lifts out the water in buckets until the amount of water becomes so much that it is no longer possible to dig deeper. If 1 bucket of water was lifted out per 10 buckets of mud, it implies that the discharge of test well is low, whereas 10 buckets of water per 1 bucket of mud herald a high discharge (Semsar Yazdi 2010: 313).

In many regions of Iran each permeable, semi permeable and impermeable layers of soil are called by a special name which gives credence to the fact that Iranian qanat masters have enjoyed a vast traditional knowledge on geological conditions.

3.3 Indigenous Pedology (Types of Soil)

The qanat masters use the physical traits of soil to define and classify it into different types. For example they make distinction between soils according to their color, porosity, odor, gas content, hardness, softness, etc.

Khak-e Solb: hard soil with low permeability (Semsar Yazdi 2010: 312).
Zardeh Shoolat: yellow clay which collapses or subsides to a large extent.
Chilar: mass of pebbles which may be found underground and their size varies from a pea to a walnut. In the water production section these pebbles benefit the qanat.
Rekhve: soft, damp and crumbling soil in which excavation is difficult (Ashasi and Safinejad 2000: 52).
Ros: ordinary clay which is compact with very low porosity and almost impermeable to water.
Rig o Ros: combination of clay and pebble with a good porosity which ease water seepage through it. This soil is suitable for qanat especially in water production section.
Ros-e-Suraxdar: clay with high porosity.
Rig-e-Badi: dune which lies beneath younger layers of sediments. This soil is very crumbling and needs to be shored up when qanat cuts through it.
Seqat: rubble stones which lack any cement or clay and can drain out qanat water especially in water transport section.
Sel: very hard stones regardless of their size, which can be found underground or on the earth surface.
Sang-e-Atishi: flint which is hard and white in color.
Sanglax: layers of ground awash with peddles and rubble stones with different size and little clay in it, which is widespread in the mountainous area of Yazd.
Sang-e-rag rag: layered stones which are considered an indication of a good discharge. The qanat masters believe that such stones can convey water from a long distance.
Shoolat: very crumbling soil in which it is not possible to tunnel without reinforcing and lining with firm materials.

3.4 Estimating Qanat Discharge

After the qanat workers finish with the test well, they estimate the qanat discharge. To do so they dig the test well into the aquifer as deep as water seepage allows (usually 50 cm into aquifer). The qanat master empties water out of the test well by a bucket, and then he waits for water to refill the well. He calculates the time it takes water to fill the same place, and then judge the possible discharge of this qanat if built. The longer it takes water to refill the well, the more discharge the qanat is going to have. As mentioned According to Haji Abbas Nasiri, a qanat master in Yazd, If 1 bucket of water was removed from the well for 10 buckets of mud, it means that the discharge of well is to be low, whereas 10 buckets of water for 1 bucket of mud indicate a high discharge (Semsar Yazdi 2010: 313).

Sometimes the qanat master measures water seepage in the test well in different seasons in order to control the effect of seasonal precipitation on water infiltration into the test well. In some aquifers the amount of water seepage closely correlates with seasonal rainfall, so one cannot expect the same water seepage in summer as in spring. If the qanat master builds a qanat in such an aquifer, the qanat water would be fluctuating, not reliable throughout the year. The discharge of such qanats culminates in spring and dramatically dwindles at the end of summer and in the beginning of fall. The discharge starts increasing as soon as the first precipitations take place in fall.

3.5 Estimating Water Quality

According to qanat masters there is no way to make sure about water quality but to dig a test well and see its water up close. Nevertheless they can conjecture about water quality of a possible qanat by looking into water of nearby streams, springs or other qanats. Water quality is associated with soil condition and types of geological formations in the site of qanat construction, so such indications can give a clue to the qanat master to guess how the qanat water would be. For example if the ground surface would be encrusted by salt, the qanat master cannot fasten his hope on a qanat to drain out pure fresh water.

Karaji explains some interesting experiments on the groundwater quality in his book as follows: "If you notice any change in the color of water, take it for granted that this water is not good, and if the water smells or tastes bad, it is certainly rotten. If you come across two sources of water which are both seemingly pure and appropriate but you are not able to distinguish the better one by examining their physical traits, you can experiment with them by two bowls. Fill two similar bowls with water from each water source and then weigh the both bowls. The lighter bowl contains the purer water. Also you can take two new potteries with the same size, texture, hardness and color. Fill up each pottery with water from each source and then hang them from a tripod. Place two glass containers each beneath a pottery to collect the water drops seeping out of the potteries and dripping down. Wait for one or 2 h and then conclude that the pottery which has exuded more water it contains the purer water" (Karaji 1966: 29) (Fig. 3.4).

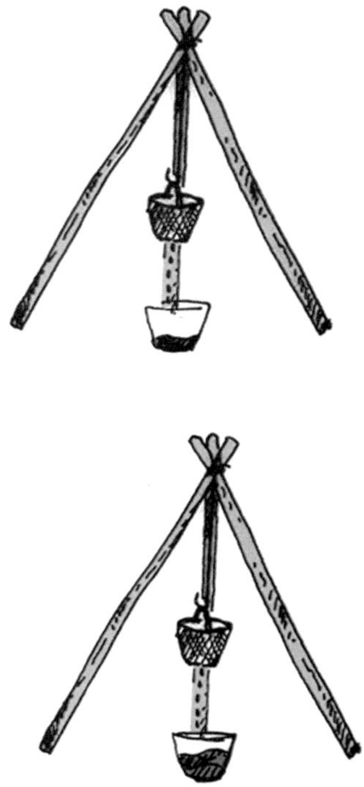

Fig. 3.4 Two potteries hanging from tripods, an experiment to find the purer water

References

Ashasi A, Safinejad J (2000) Qanat glossary. Iranian ministry of energy. Yazd Regional Water Authority, Yazd

Bonine Micheal E (1982) From qanat to kort: traditional irrigation terminology and practices in central Iran. IRAN J Br Inst Persian Stud XX:145

Ellis AJ (1917) The divining rod a history of water witching. Department of the Interior. United States Geological Survey. Government Printing Office, Washington

Maryam H, Hossein A, Mahmood K (2014) Use of desirability function method in optimization of regeneration and callus induction of Alhagi camelorum. Am J Plant Sci 5:268–274

Karaji Abu Bakr ibn Muhammad ibn al Hassan (1966) Extraction of hidden waters. Translated from Arabic to Persian by Hossein Khadiv Jam. Tehran: Bonyad Farhang Iran Publication

Semsar Yazdi AA (2010) Qanat from practitioners' point of view. Iran Water Resources Management Company, Tehran

Taubert Paul Hermann Wilhelm (1981) Illustrated handbook of succulent plants. Leguminosae. In: Engelmann (ed) Natürliche Pflanzenfamilien, vol. III, 3. Springer, Berlin

Chapter 4
Digging of Shaft Wells and Qanat Gallery

Abstract This chapter encompasses one of the most important issues on qanat construction which is how to shape a qanat in a traditional way. The qanat masters cope with different conditions through which the qanat gallery cuts. They have to figure out how to continue building qanat whether in a hard geological formation or in a soft crumbling one. A qanat system may be as long as 70 km, accordingly the qanat workers may come across a verity of obstacles or problems to which they have invented effective techniques. This chapter starts with leveling techniques and covers such issues as well boring methods, digging of water production and transport sections, sinking shaft wells into aquifer, intersecting qanats, etc.

Keywords Leveling • Tunnel digging • Water transport section • Water production section • Underground navigation

4.1 Introduction

As mentioned, after the qanat construction site is located, a test well is dug in order to find out about water quality and quantity. The test well is dug some 20–50 cm into water saturated area underground and then the workers stop digging deeper and turn to dig the horizontal tunnel. Next step is to dig water transport section of the gallery, but first of all they have to do some calculations in order to designate the exit point of qanat. It is very important to estimate where the exit point is going to turn up, because water flows down according to gravity and it should not appear on the surface down slope from the farmlands. After they pinpoint the qanat appearance, they look into the soil down there to make sure about its fertility, and then they get started on the construction of qanat. In case the soil turns out unsuitable for cultivation, the workers give up digging the qanat and manage to find a better place. Therefore leveling is done to specify the position of the farmlands on one hand, and to properly build the qanat tunnel on the other hand. By leveling the workers can know how long the qanat is going to be. If the qanat's length is more than expected, it may do not make economic sense to keep working on it, if its cost is more than what they afford.

A.A. Semsar Yazdi, M. Labbaf Khaneiki, *Qanat Knowledge*,
DOI 10.1007/978-94-024-0957-4_4

In fact they calculate the high differences between many spots from the test well down. Thus they can know how deep each shaft well should be, so that the gallery connecting all the shaft wells would be almost horizontal. The qanat masters believe that leveling is necessary in water transport section, but it is not needed in water production section, because in there water flow leads the workers to observe a proper gradient along the tunnel. If the tunnel would be dug too steep water would rush down, and if it would be too gentle water would become stagnant. When it comes to water transport section, it would be much more difficult to dig a long tunnel with an optimum gradient all the way where water is absent. Therefore they have invented the method of leveling to keep the desired gradient in water transport section.

4.2 Determining the Direction of Qanat Gallery

Another important issue in qanat construction is how to determine the direction of qanat. After the beginning (test well) and the end (exit point) of a qanat came to light, it is the turn of the direction of gallery to be specified. It seems more logical to connect the qanat beginning and end together with a straight line. To do so, on the surface the workers mark the places where next wells are to be dug one after another, by piling up some soil and sand. These piles show the direction of the new qanat on the ground surface from the test well down to the exit point. In Yazd region these piles or marks are called “Korom”. The qanat masters place the marks in a straight direction, because a direct tunnel means less excavation and less cost unless they come across an obstacle underground which get them to change their direction. After marking, they begin leveling on the ground surface. Methods of leveling differ in different areas, and we can mention four types of leveling according to different topographical conditions and earth relief, as follows: simple leveling, double leveling, leveling in hilly lands and leveling between two parallel qanats.

4.3 Simple Leveling

This leveling is the simplest one whose principles apply to the other types of leveling. Leveling is referred to as calculating the gradient by means of a traditional instrument named “Taraz”, though in the past decades another tool has also been introduced. This tool is a plastic transparent hose in which the workers put some water. On a horizontal surface the hose would be horizontal too, and no water would pour out of it. Nevertheless most of the qanat masters still prefer to use the same traditional “Taraz” which is a simple tool consisting of two 1.5 m wooden poles with a long thin rope in between. The rope is about 30 m long, tied to the top of both poles. The poles are held upright on the ground in some distance from each other. To make sure that the poles stand completely upright, they cut a rectangular hole in each pole and hang a plummet from the top of hole. The plummet position implies whether or not the pole is completely vertical (Figs. 4.1 and 4.2).

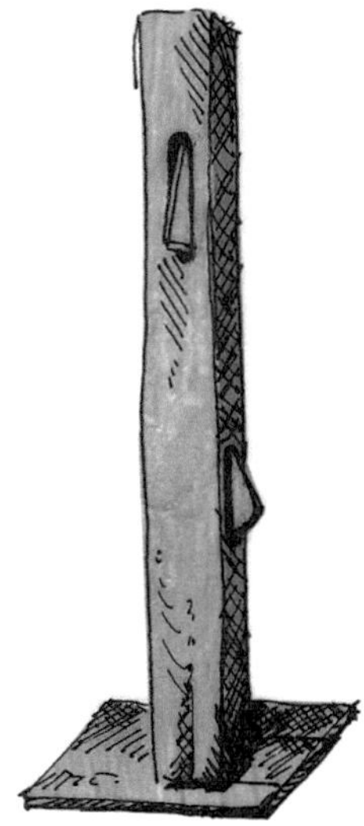

Fig. 4.1 The pole of the leveling tool

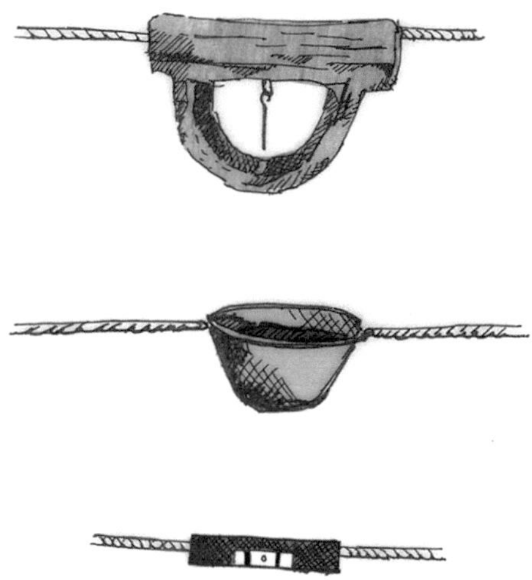

Fig. 4.2 Different types of level

To know if the rope between the two poles lies completely horizontal they use a leveling tool which consist a small glass container full of water with an air bubble in it. They put the glass container gently on the role and consider the rope horizontal if the bubble stays exactly in the middle. In the past instead of this glass container, they used a small rectangle of wood with four small holes exactly in a line. They used to attach the rectangle to the middle of the rope and hold a plummet (hanging from a string) on it, just behind the four holes. The rope was considered horizontal if the plummet string crossed all the four holes exactly in the middle.

As mentioned earlier a worker holds a pole upright on the ground surface close to the test well and the other worker walks downstream and holds another pole exactly where the second well is to be sunk. The two workers hold the poles some 15–30 m away from each other, and this distance depends on surface gradient, soil type, ground reliefs, etc. Another worker places a level tool on the rope stretching between the two poles. Afterward he tells the worker standing up-slope how much to lower the rope along the pole until the rope lies completely horizontal and the bubble stands in the middle of the level tool. The depth difference between the test well and the other well downstream is the same length the rope is lowered along the pole in order to get the rope horizontal. For example if the rope comes down the pole one meter, it means that the next well should be one meter shallower than the test well. In this case the tunnel connecting the bottoms of these two wells lies horizontal with no slope. But in reality every qanat tunnel enjoys a gentle slope which allows water to flow down. The qanat masters dig the downstream wells slightly deeper than what supposed to be. For example if the next well is to be one meter shallower than the test well, they dig it 95 cm shallower in order to make a gentle slope along the tunnel. There may be some small ups and downs along the tunnel, which would be leveled after water starts flowing and shows every unevenness on its way.

In plain areas the qanat practitioners tries to observe the tunnel slope as gentle as possible in order to make the tunnel shorter and convey the water sooner. In the plain areas, the shorter the tunnel, the less money they have to spend on tunnel excavation. For example in a land with a surface gradient of 0.3 %, the tunnel cannot run with the same gradient, because parallel lines never intersect. In such areas the qanat masters envision a very gentle slope like 0.1 % for the tunnel in which water can flow only very slowly. The locals call such a slow current "Mat". They believe that even this almost zero slope is enough to drive the water out of the tunnel, because of water weight which pushes the current out. For example in Bam, Kerman province the wells are dug a little more than what the leveling tool suggests, so that the tunnel would eventually run at a general slope of 0.2 %.

In case of mountainous qanats the situation is quite the reverse. In such areas the surface is so steep and a tunnel with extremely gentle slope would appear on the surface so shortly. Such a qanat may appears up slope from the cultivated lands or may bring about other technical problems, so the qanat masters envision a steeper slope taking into account the location of water consumption and soil vulnerability to water erosion (Fig. 4.3).

4.4 Double Leveling

In the steep mountainous areas the workers sometimes have to repeat leveling several times only between two wells. In other words the worker A stands by the test well and the worker B holds the pole just on the location of the next well. The pole is only 1.5 m long and the rope may not be horizontal even though the rope is lowered to the end of the rope. In this case the workers have to stand closer and do serial measuring between the two wells and then sum up the numbers. For example if the rope comes down 1.5 m along the pole in the first leveling, and then comes down 1 m in the second leveling, the next well should be 2 m shallower than the test well. Thus the tunnel runs almost horizontal beneath these two well (Fig. 4.4).

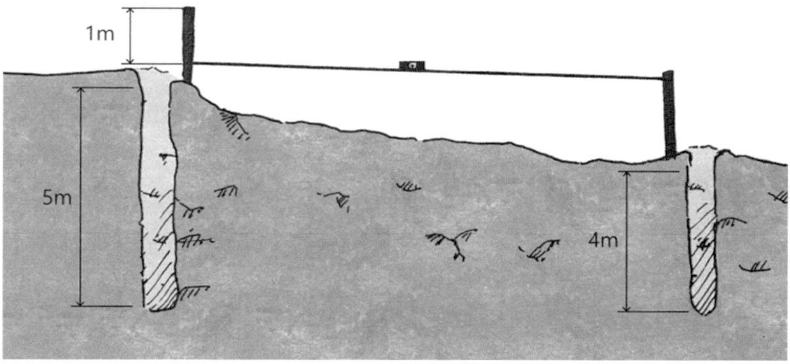

Fig. 4.3 Simple leveling

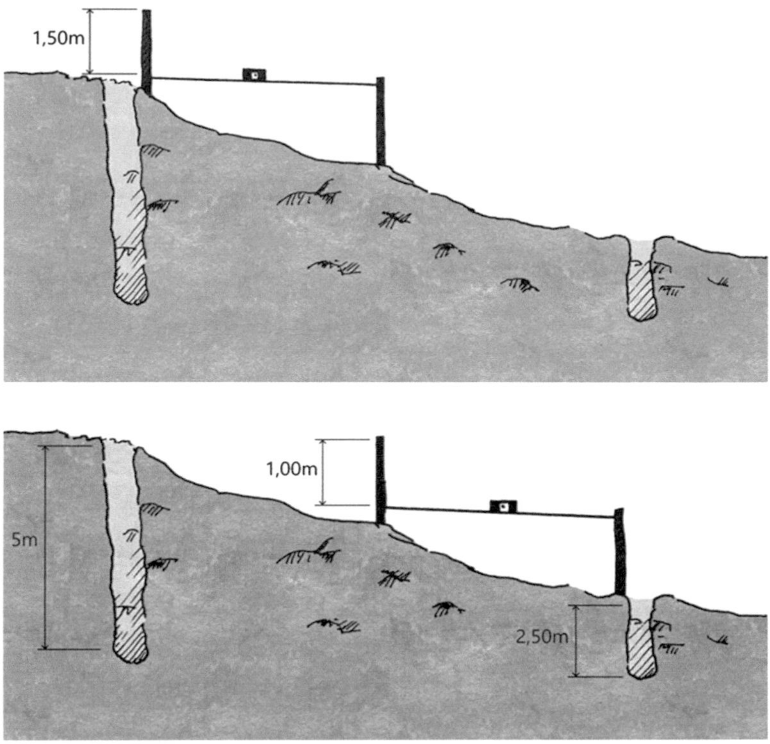

Fig. 4.4 Two stages of double leveling

4.5 Leveling in Hilly Lands

Another problem that may come about is how to use leveling tool in undulating lands with lots of ups and downs where there may be positive and negative gradients at the same time. In case there would be a bulge between two wells, the qanat masters have to do leveling twice. They measure the height difference between the test well opening and top of the bulge and then between there and the next well. Afterward these two measurements are subtracted so their difference is how much the next well should be shallower than the test well up stream (Fig. 4.5).

For example worker (A) stands just close to the test well holding a pole. Worker (B) goes up the mound and then gets the rope down the pole until it lies horizontally. They make a note of how long the rope comes down the pole, and then go on measuring. Afterward worker (A) walks up the mound and worker (B) walks to the location of next well, and they measure the high difference again and make a note of the number. They subtract the two numbers whose difference shows how much the next well should be shallower than the test well. The standard image of qanat is that its shaft wells constantly get shallower as we approach the qanat exit, but some-

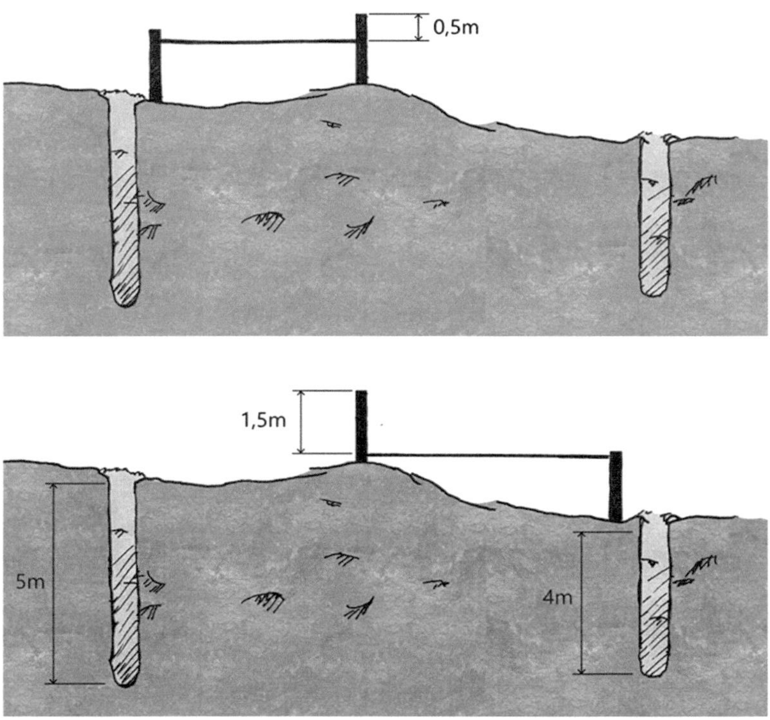

Fig. 4.5 Two stages of leveling in hilly lands

times this image does not apply to the qanats in hilly lands. The qanats which run beneath hilly lands may have sort of anomy in terms of depth of shaft wells, though their tunnel slope remains unchanging.

4.6 Leveling Between Two Parallel Qanats

This leveling is practiced where the qanat bound is to be determined. As an instance in the region of Bam there are 23 qanats running parallel west east just a short distance from each other, crossing the well-known Bam fault. If the locals want to build a new qanat in between the existing ones, the new qanat is not permitted to be dug deeper than the others. Therefore the depth of the surrounding qanats should be observed, while digging a new qanat in order to prevent any interference between their discharges. They place the leveling pole just beside the first well of the new qanat and start measuring the high difference between this point and one of wells of the nearby qanat just across from this point. Thus the permitted depth of the new well can be calculated. In fact the first well of the new qanat and the nearest well of the older qanat should be of the same depth. For example if the well of the older qanat would be 30 m deep and the height difference would be 1 m, the new well is expected to be 29 m deep (Fig. 4.6).

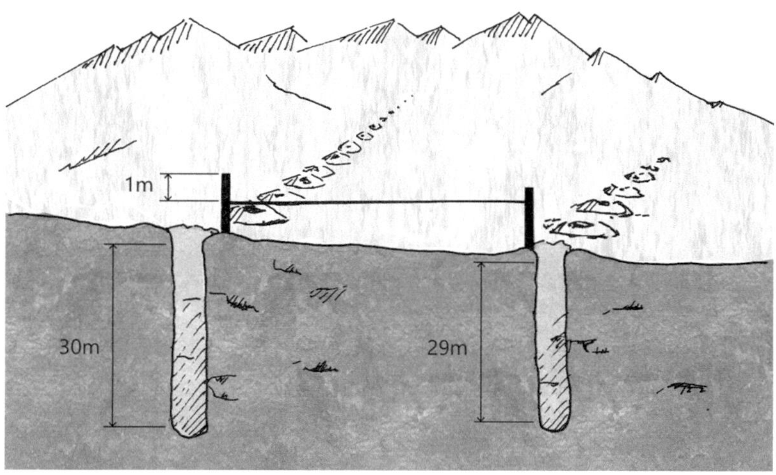

Fig. 4.6 Leveling between two parallel qanats

4.7 Qanat Construction

After leveling is accomplished, the workers get started on qanat construction. Leveling is carried out from upstream to downstream, but digging takes place the other way around; from downstream to upstream. Water transport section is the first part of a qanat to be built in order to provide a way through which water can be drained out. Afterwards water production section is built to tap the groundwater as described below.

4.8 Digging Water Transport Section

In general from the exit point to where groundwater starts dripping from the tunnel sides and ceiling is referred to as water transport section and from this point up to mother well is called water production section. A well located between the WTS and WPS is called Taroon-Khoshkoon which is a mark identifying the border between water bearing and dry zones (Fig. 4.7).

After leveling is done, the shaft wells are dug and then the wells are connected at their bottoms one after another. This tunnel forms the main part of the qanat's structure. Test well is the first well which is dug to find out about the quality and quantity of groundwater. After sinking a test well, the workers manage to dig WTS in order to provide a way to drain the water which later seeps into the tunnel and builds up there, otherwise it is impossible for the workers to advance into the aquifer for more discharge. Usually the next well dug immediately after the test well is called "Jofte Badoo" which locates in WTS, some distance downslope from the test well where there would not be any water infiltration. Jofte Badoo is actually a twin well consisting of two wells dug some 1.5 or 2 m away from each other to better ventilate the wells especially where soil emits harmful gas. Some of the geological formation gives off poisonous gases that are obviously harmful to the workers, so the workers dig two wells simultaneously and connect them every 10 m by a short horizontal tunnel in order to ease air circulation as they go down. In this case fresh air comes down from a well and goes up through the other. In most cases the workers prefer to

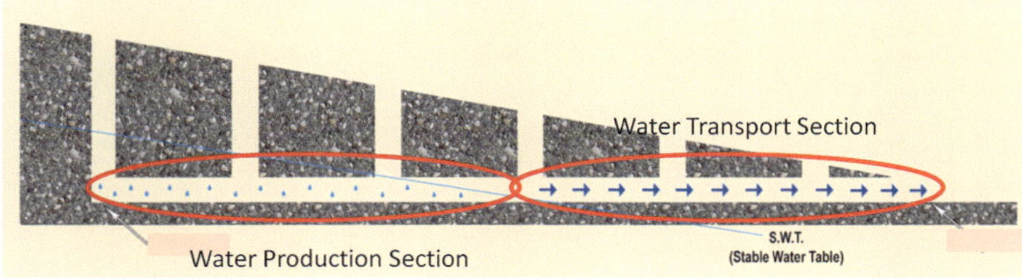

Fig. 4.7 Qanat profile with water production and transport sections

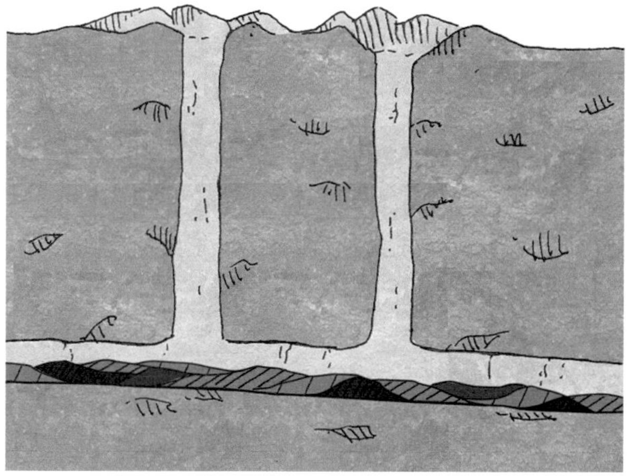

Fig. 4.8 Jofte Badoo, a twin well to better ventilate the wells

dig twin wells for some other purposes as well, which are mentioned later. After the twin wells are done, their bottoms are connected by a tunnel and then they continue digging toward the test well, while another worker digs at the same time in the opposite direction toward the exit point. Two windlasses simultaneously operate above the twin wells to halt the excavated soil onto the earth surface and two qanat masters dig the tunnel in the opposite direction. Thus another application of twin wells is to speed up qanat construction. As a result the applications of twin wells can be summarized as follows (Fig. 4.8):

1. Ventilating the wells and tunnel
2. Easing extraction of soil and debris
3. Accelerating construction of tunnel

In WTS, wells are dug firstly at the depth already calculated by the leveling tool. After a well is finished, two workers start digging tunnel from the well bottom in both opposite directions. As mentioned the twin wells get connected to the test well up slope and also to another well down slope. Afterward another well is dug down slope and then the two workers start digging the tunnel at its bottom in two directions. This procedure continues until the qanat exit point. There is a common question on which part of a typical qanat would be built first. Needless to say WTS should be built first of all in order to provide a way to drain out groundwater, otherwise water build-up stops the workers from digging further through water saturated area underground. In a typical qanat, the workers start digging the WTS some distance down slope from the test well which indicates the border between WTS and WPS. Two groups of workers dig this tunnel in two directions to speed up their work; toward the test well and the exit point at the same time. When they wrap up WTS, they can get started on the WPS.

In the region of Bam, the qanat tunnel is dug as an open trench where the tunnel approaches the exit point and its depth does not exceed 4 m. This open canal is called "Kush" which gets shallower and shallower as approaches the earth surface, and then it is called "Ju" which means ditch. In WTP the tunnel ceiling is built angled and rectangular which is called "Sandoghi" or box-shaped.

4.9 Digging Water Production Section

As mentioned WPS can be built only after WTS is completed in order to drain out the water seeping into the gallery. As the workers dig deeper into water bearing zone, water seepage into the tunnel increases and eventually water can flow down the tunnel toward the exit point. At this stage the workers do not usually content themselves with the small temporary springs of water that may show up while they are digging the WPS, but they prefer to continue until the tunnel hits a reliable source of groundwater.[1]

If the tunnel cuts through water saturated area, water infiltration is much enough to pave its own way toward the exit point and travel the WTS, otherwise the workers have to do something to ease the water flow which is called "Ab Pishbari". As mentioned the tunnel slope is so gentle that water needs to be moved forward by a shovel. Thus a qanat master follows the water current along the tunnel and explores the tunnel all the way to remove any possible barriers on the way of water. He dumps the materials removed from the tunnel in a bucket being pulled up by someone else working with the windlass on the surface. As the worker bends and walks toward the exit point and keep pace with the water flowing down, the windlass is moved from one well to another until the exit point. In the meantime the tunnel is extended through aquifer and water grows more and more. In most parts of Iran, the tunnel end in WPS is called "Pishkar" which has the most penetration into the aquifer. The further Pishkar goes into the aquifer, the more water seeps into the tunnel. The workers try to observe the same gradient as that of WTS when they extend the tunnel into aquifer to drain more water, while another team of workers dig new shaft wells upstream in advance. It is very rare for the workers to make mistake in terms of tunnel slope in WPS, because they consider the slope such that water level always remains knee high while advancing into aquifer. It implies that there is something wrong with the tunnel slope if water level deviates from what expected. In WPS the worker dig the ceiling and sides of tunnel first and then it comes to the tunnel floor. In plain qanats water seepage usually take place in sides of tunnel, whereas in mountainous qanats it is possible to face water seepage even in the ceiling. Such qanats are called "raining qanats" whose tunnel ceiling should be dug like a pitched roof to prevent water drops from falling down directly on the worker's head. In Iran

[1] In Bam region sometimes the workers come across temporary springs while digging the tunnel, which are called "Benow". Benow is actually small reserves of water trapped in underground cavities, which spring up in the tunnel and flow down. According to the qanat masters there may be a considerable amount of water gushing out from the ground into the gallery, whose temperature is less or more than that of the normal current of water already flowing in the tunnel. The workers do not count on these springs which dwindle after a while.

such ceilings are called "Makooyi" or "Mehrabi" whose construction demands a great deal of expertise and experience. Such ceilings not only make water drops slide down the steep planes, but also reinforce the tunnel and reduce the risk of collapse. After the worker digs the tunnel ceiling and sides, he turns to the floor to the extent that water would not come higher than his knee level (Fig. 4.9).

In Bafgh[2] region if there would be no water in the tunnel or water would not be enough to make a current and it would be impossible to observe tunnel slope using water level, they resort to using light instead. To do so the worker places a lamp on the tunnel floor and walks some distance up slope from it and then puts his head on the floor and tries to see the light. If the light is hardly visible, it means that the tunnel is steeper than needed. If the light is elevated, it implies that there is a reverse slope in the tunnel, which should be corrected. According to traditional qanat masters if the tunnel slope would be too steep, water would rush down the tunnel and wash away some parts of tunnel, causing erosion. Moreover a steeper tunnel means a longer qanat which consumes more money and time to be built. On the other hand a tunnel which enjoys an extremely gentle slope would hamper water flow and make water almost stagnant, bringing about sedimentation and eventually obstruction of tunnel.

Fig. 4.9 In WPS the worker dig the ceiling and sides of tunnel and then the tunnel floor, he watches the water level to remain knee high

[2] Bafgh is a town in and the capital of Bafgh County, Yazd Province, Iran.

To dig WPS it is common to use a special pick axe called "Abgir Kani" whose point is about 70 cm long. In case water level would be below the worker's knee, it turns out that somewhere in the tunnel has a reverse slope not letting water flow down freely. So the worker looks for this reverse slope and fixes it with the help of the aforesaid pick axe. Sometimes a part of tunnel collapses and leaves an obstruction, making water build up in the tunnel. In this case water level goes up and makes the worker mix up this accident with a reverse slope. The worker may be misled by a collapse-induced obstruction and accordingly he may decide to build the tunnel steeper. In this case such a steep slope may play havoc with the structure of the tunnel. To prevent such a mistake, the worker digs a small hole into the tunnel side just after he finishes his daily work. The hole is some 10–15 cm deep just above the water level in the tunnel. This hole is called "Kanak" in the workers' terminology, which serves as an alarm if there would be any obstruction in the tunnel. The day after, if water reaches the hole, it means that a collapse has obstructed somewhere in the tunnel, making water build up and go up to that point. In this case before resuming their work, the workers manage to search for the obstruction and remove it to ease the water flow (Fig. 4.10).

WPS is extended as long as no natural or economic obstacle hinders the work. In case the tunnel hits a hard impermeable ground, the workers have to give up extending tunnel. Also economic problems can bring the extending to a standstill. For example if the qanat shareholders no longer afford to pay for the extending and there would not be sufficient budget or the tunnel enters the vicinity of other qanats, the extending would be stopped. Sometimes the water is so much that the WTS gets eroded and damaged and extending is no longer possible.

Here it is worth noting the dimensions of tunnel in a typical qanat. Qanat tunnel in WTS is usually 1–1.1 m high and in WPS is at least 1.5 m. Needless to say WPS should be higher to increase the seepage surface inside the tunnel. A typical tunnel is between 60 and 70 cm wide. The workers use their forearms to measure the width of tunnel. In general in a tunnel cross section the lower part is wider than the upper one in order to increase the infiltration capacity in lower part of tunnel which is exposed to saturated ground the most. Moreover the upper part is envisioned narrower to avoid unnecessary digging and reduce the project cost and the risk of collapse. Another factor which affects the tunnel height is the distance between the shaft wells. The farther the wells, the higher the tunnel in order to better facilitate air circulation and ventilation in the tunnel (Fig. 4.11).

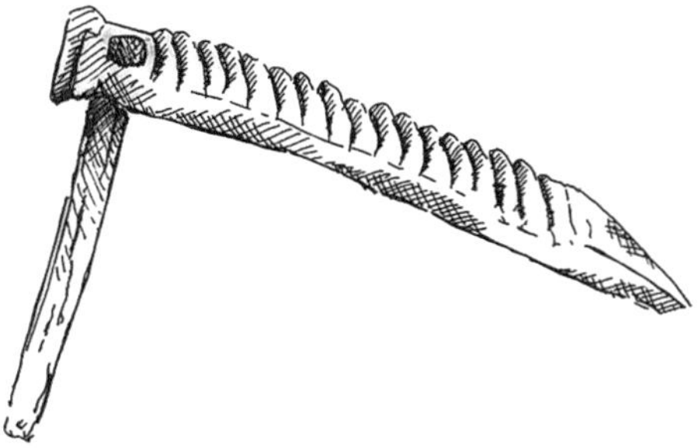

Fig. 4.10 "Abgir Kani" pick axe

Fig. 4.11 The workers use their forearms to measure the width of tunnel

4.10 Distance Between Shaft Wells (Poshte)

The distance between two shaft wells is called "Poshte" in central Iran, which depends on depth of shaft wells as well as type of soil in which the wells have been sunk. This distance is usually twice the depth of wells, though it completely depends on qanat master's judgment and may vary with types of soil. For example if the gallery would be so stuffy and the workers would have problem with breathing, they envision a shorter distance between wells in order to have better ventilation, though nowadays new technologies can be used to ventilate the tunnel at lower cost. Also, the wells are dug closer if the tunnel cuts through soft and crumbling soil. In the crumbling soils the tunnel should be shored up with concrete hoops which are carried to the tunnel through the shaft wells. The closer the wells, the easier it would be to carry the hoops and put them in place. In contrast the wells are dug relatively far away in case the qanat cuts through hard geological formation. In fact sort of economic equation dictates how far the wells would be from each other. When the conditions are not favorable for digging wells the workers dig as few wells as possible, for example when the soil is hard or tunnel is so deep. That is why the wells are dug more disperse as tunnel goes deeper toward the mother well (Fig. 4.12).

4.11 Shaft Well; Digging Method and Dimensions

A typical qanat well has a circular cross section with a diameter just enough for a worker digging down its bottom with a short-handle pick axe. First of all, the worker puts his pick axe on the ground surface the way the pick axe can be rotated around its point and makes a circle on the soil with the end of its handle, like compasses.

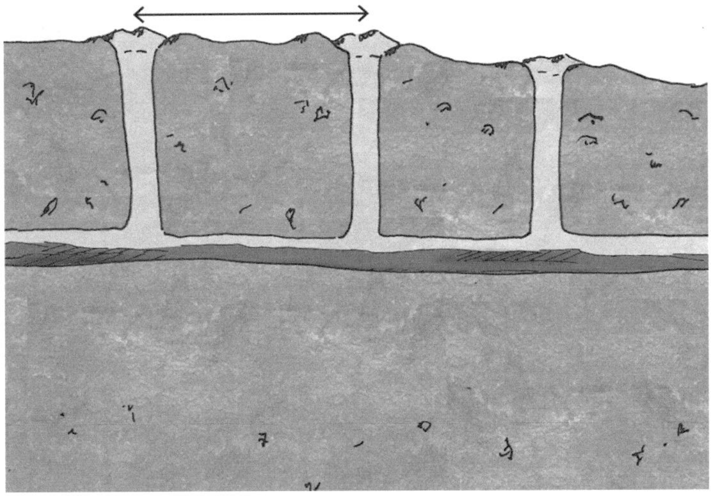

Fig. 4.12 "Poshte"; the distance between two shaft wells

The handle is usually 35 cm long, so the circle would be some 70 cm in diameter. Afterward the worker starts digging inside this circle. In fact the well diameter is decided by the qanat master and should be wide enough for the workers and buckets to go up and down without touching the wall sides doing damage to the well. On the other hand the well should not be wider than enough to prevent unnecessary cost and labor. In general the well width should be suitable for workers and materials transportation as well as ventilation. To do so, the optimum well is considered to be 70–80 cm wide, but it may be reduced to 60 cm to speed up the work if the qanat master decides. In case the soil is so soft and crumbling that the well needs to be lined with concrete hoops, the well is envisioned wider, even up to 90–100 cm (Figs. 4.13, 4.14 and 4.15).

Traditional qanat workers crouch down inside the well, while digging. They dig the well in a clockwise spiral way as if they climb down a tower stairs. Therefore one side of the well bottom is some 30–40 elevated than the other, which is locally called "Passa". A skilful worker is a worker who can observe "Passa" properly. According to the qanat masters in Taft, in distant past the qanat practitioners did not have sufficient expertise to dig circular wells with observing Passa, so they resorted to digging rectangular wells instead. The locals call such rectangular wells "Kermani" or "Gowri" wells.[3] They believe these wells have been dug by Zoroastrians long ago before the advent of Islam. The dimensions of such wells are usually 80 by 120 cm, which can be found across Iran (Figs. 4.16 and 4.17).

[3] The word Gowr is a local pronunciation of the Persian word Gabr which was introduced to the ancient Persian from Aramaic language. In Aramaic language this word appears as Gabra which means man or male. When this word entered Persian it was used only to refer to Zoroastrians. Before the advent of Islam Gabr had a good or at least neutral connotation in Persian language, but during the Islamic period this word was referred to someone involved in heresy. This word has repeatedly been used during 1300 years history of Persian literature.

Fig. 4.13 Using a pick axe to determine the well radius

Fig. 4.14 Well shored up with concrete hoops placed on four vertical hoops

Fig. 4.15 Well shored up with concrete hoops fixed on the well sides

Fig. 4.16 A well is dug in a clockwise spiral way

Fig. 4.17 Rectangular well

Here it is worth noting that another reason why the ancient workers managed to dig such wells was to better find the right direction underground while digging a tunnel between two wells. In a deep circular well it is very difficult to figure out which direction leads to the next well, while digging the tunnel. The past workers used to dig rectangular wells probably to make up for the lack of navigation tools, because such wells were all dug in one direction in a row. When they dug into the width of a rectangular well straight ahead, it eventually led to the width of the next well, though they had to dig and move a larger mass of soil. They could get rid of such a labor later by inventing navigation tools which let them dig circular wells. Thus the qanat practitioners came to invent some devices to navigate from and to a well underground. One of these navigation tools is called "Rassi" which will be described later.

Moreover in many Iranian qanats there are rectangular wells which have nothing to do with ancient methods. Some rectangular wells have been dug recently in order to tackle the problem of digging into water saturated soil, when it is impossible to dig a well below the water table. In this case they have to dig such wells from the tunnel upward so that water can pour down and flow in the tunnel toward the qanat exit. This well is called "Devil" which will be described in details. Such a well should have a rectangular cross section to let the worker dig overhead without being hit by the materials falling down. The worker digs half the rectangle while standing safely under the other half.

4.12 Ascending Tunnel

Sometimes when the qanat workers dig forward in the gallery into the aquifer, they may come across a hard and rocky layer into which it is very difficult to continue digging the tunnel. In this case the worker tries to avoid such a formation as much as possible. They prefer to bypass or even go over such a layer. Going over a hard layer and digging the tunnel with the same slope as that of the layer is called "Roo-Neshastan" or ascending tunnel. In this case both tunnel and hard layer runs parallel toward the mother well. For example in the qanat of Chahok in Taft, Yazd there is a side branch which is 200 m long, running over a rocky layer with a relatively steep slope. In this case as we approach the mother well, the shaft wells get shallower rather than deeper as seen in a typical qanat, because the tunnel is steeper than the earth surface (Fig. 4.18).

4.13 Underground Navigation; Building Tunnel Between Two Wells

Regardless of which section of qanat is to be built, a good underground navigation is of great importance, because the tunnel should run exactly beneath all the wells, connecting them to each other. Any mistake or deviation in the tunnel direction takes a toll on the qanat owners from different points of view. So the qanat master tries his best to dig as short tunnel between two wells as possible. To avoid turning from the right direction and getting lost underground, the qanat workers have invented a simple tool named "Rassi" which dramatically reduces the risk of mistake. Rassi consists of a wooden or iron stick 100–120 cm long with two strings tied to it, some 20–30 cm away from each other (Fig. 4.19).

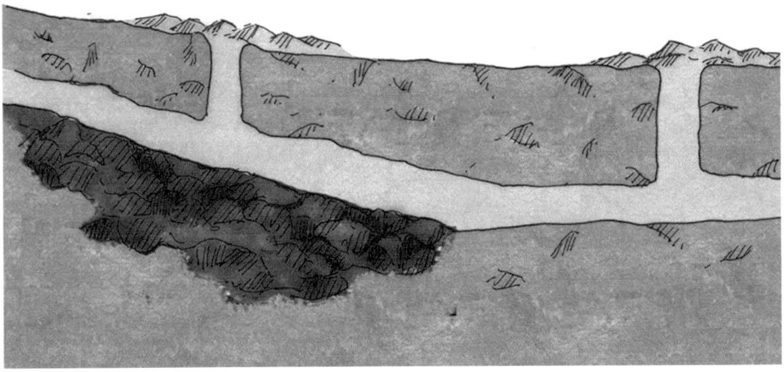

Fig. 4.18 Ascending tunnel

Fig. 4.19 Rassi for finding the right direction underground

Two stones hang from the aforesaid strings, forming kind of plumb-bob. The qanat master places the stick horizontally on the mouth of the well A, such that the stick exactly points to the well B, the well to which the tunnel is to go. Now the stick points to the right direction and down the well the stones hanging from the stick point to the same direction. The worker holds a lamp just behind the two stones the way that one stone eclipses the other and only one shadow would be shed on the opposite wall. This shadow tells the worker where to dig toward the well B. This spot on the side of well where digging starts is called "Sineye Kar" literally "work front". In order to make sure that the tunnel has remained in the correct direction, once in a while – usually every 5–10 m – the worker puts a lamp at the starting point and observes it carefully. If the lamp is still seen in the middle, the tunnel is considered going in the desired direction, otherwise it should be corrected to left or right according to the light. The ancient qanat workers knew by instinct that light always travels in a straight line.[4] In the past the qanat workers exhibited a wonderful accuracy just by using such simple tools. For example in the qanat of Khaje Askar in Bam some parts of its tunnel have been built long ago by traditional tools. Where someone holds a lamp in its tunnel, the light is still visible from afar sometimes one kilometer away. The tunnel is incredibly straight.

In the wake of the introduction of the magnetic compass, it has been abolished to use Rassi, and it has been replaced with normal compass. Nevertheless in some cases the workers cannot show the same accuracy as they did in the past. At present they put a compass on the ground just close to the well A with the N mark of the compass pointing to the well B. In this position the compass needle would be fixed on a number which should be kept in mind. Now the worker takes the compass to the bottom of the well A and adjusts the compass so that its needle would be set on the same number. Now the mark N points to the right direction where the worker gets started on digging (Fig. 4.20).

In reality some of the workers are illiterate and almost unable to read the compass numbers, and in fact they memorize the approximate position of the compass needle. They may mix up one number with another, which leads to a deviation from the correct direction. In this case the tunnel does not reach the well already prepared, and it passes by the well. If they dig longer than the surface distance between the two wells, it turns out that a mistake has been made and they have to get back to the correct direction. In this case they have to dig a lateral tunnel toward the missed well, and this tunnel may be several meters long.

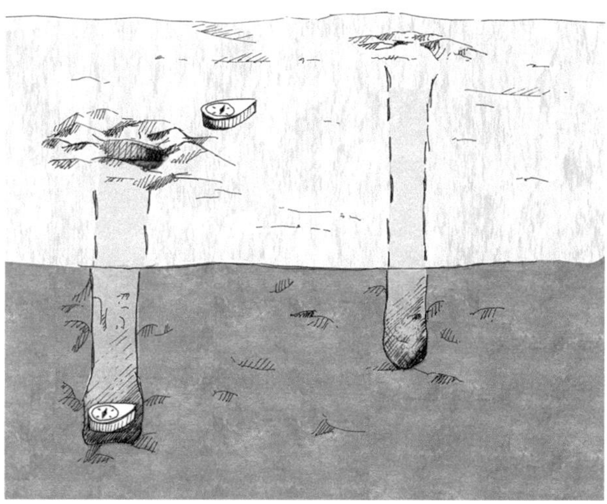

Fig. 4.20 Using magnetic compass in qanat construction

[4] In about 300 BC, Euclid wrote Optica, in which he studied the properties of light. Euclid postulated that light travelled in straight lines (www.wikipedia.org).

For example if two wells would be 50 m away from each other, the tunnel between them is expected to be almost the same 50 m, otherwise the tunnel has gone awry. In this case a tunnel is dug from the second well toward the deviated tunnel in the way that the tunnel goes forward but the well remains apart from this new direction. In Bam region this well is referred to as "Pistooyi" which means "off the beaten path" or out of main direction. Over the past years the number of such wells has been on the rise, which may have something to do with using compass instead of traditional tools (Figs. 4.21 and 4.22).

Fig. 4.21 Using Rassi to navigate underground

Fig. 4.22 "Pistooyi" well

4.14 Reverse Digging of Well: "Devil"

As mentioned some workers dig 2 or 3 wells in advance while some other workers are busy digging the tunnel, so that the tunnel reaches the wells. But in saturated area it is not possible to dig the wells deeper than the water table, because water level comes up and drowns the worker. Therefore the workers give up digging a well as soon as it reaches the water table, and then they go about digging the next well in the hope that groundwater would be drained out by the qanat gallery in the meantime and water table goes down. If no considerable drawdown occurs in water table, the workers manage to dig the rest of the well from the tunnel upward. This practice is called "Devil" in Persian. In many regions no considerable change occurs in water table, though the tunnel cuts through water bearing zone and drains groundwater out as it goes forward. In this case water build-up at the well bottom does not allow the worker to continue digging, so the worker gets back to tunnel and digs the same spot overhead until he reaches the same well. Water pours down and flows away as the worker digs overhead. "Devil" is rectangular in shape with a dimension of 80 by 120 cm.[5] That is why the worker can place on one side and dig the other side without being hit by the soil and mud falling down. The worker digs both sides of that rectangle alternatively step-wise, changing his position continuously from side to side, while sitting on a wooden scaffold horizontally braced on the sides of well. To do so on the sides of well they cut two holes across from each other to hold a wooden scaffold horizontally, so that the worker can perch on it. The worker cuts new holes and gets up the scaffold as he digs upward (Fig. 4.23).

To locate Devil in the gallery, they measure the surface distance between the last well and the well for which Devil is to be done, and then they consider the same distance in the tunnel to find the place. For example Devil would be dug 40 m from the last well, if the distance between the two wells is 40 m, using a string and level tool with the string at horizontal position. The traditional qanat workers consider the surface gradient while calculating the surface distance between two wells otherwise they would be mistaken.

Take A for the last well and C for the well on which Devil is to be done. If the worker measures the surface distance between A and C and envisages the same distance underground, the Devil would not hit the C point but it would turn up somewhere farther, because the surface gradient has not been factored in. First of all, the qanat master calculates how deep the well would be after its Devil would be finished, using level tool. Having the depth of this well, they use a simple equation to know the distance between B and D (between the last well and the spot where Devil is to be dug). This equation is very helpful when the two wells are relatively far away or there are many ups and downs in between, which make it difficult to use the string and level tool. According to the picture below, if the last well (AB) would be

[5] The shape of "devil" may make us mistake it for "Gabri" wells the ancient rectangular wells believed to be dug by Zoroastrians long ago. Note that the upper part of "Devil" is circular for it has been dug from surface down.

Fig. 4.23 Reverse digging of well: "Devil"

subtracted from the undone well or the well for which Devil should be dug (DC), and then the difference would be squared and the result would be subtracted from the square root of the surface distance between the two wells (AC) we get a number whose square root is the underground distance between the two well (BD). In other words in this way we can calculate the Devil spot underground with minimum error.

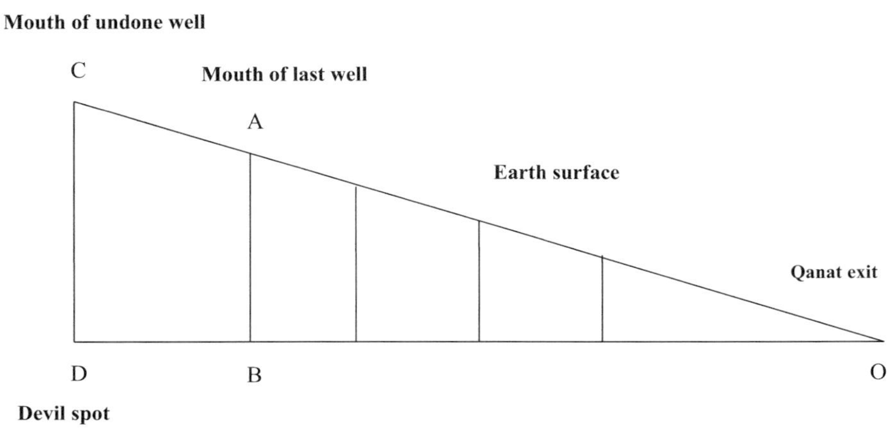

$$BD = \sqrt{AC^2 - (DC - AB)^2}$$

If we have BD, we can calculate the total length of qanat from its exit to last well by another equation on condition that the tunnel would be strait all the way. According to the following equation, if DC multiplied by BD and the result is divided by the difference between DC and AB, we get a number which equals the distance between the qanat exit point and its last well (OD).

$$OD = \frac{DC.BD}{DC - AB}$$

During our field studies we realized that at least Haj Ahmad Zahedi knew of the above mentioned mathematical equations but he talked about them in a different language. When we tried to translate them into these standard mathematical equations, we were amazed at the fact that an illiterate traditional qanat master was using Pythagorean theorem in practice. According to him, many of the qanat masters perceive such equations intuitionally and hand them down to their apprentices.

Any way the qanat master designates the Devil spot and gets started on digging overhead, but he also uses another technique named "Terek" or "Tavar" to make sure he is on track. Terek means a sound heard underground in the tunnel when someone ties a rope around a stone and drops it down the half-dug well and pulls it up and does it again. This sound leads the worker just to the bottom of the half-dug well and the Devil hits the well with no deviation. Nevertheless sometimes the worker misestimates the Devil spot, which have something to do with the presence of impassable barriers underground or lack of adequate expertise. In this case, the workers resort to digging "Bizesh".

4.15 Bizesh

In some cases the worker can pinpoint the Devil spot accurately and even starts digging overhead but he may come across an impassable obstacle like a boulder. Now it is very difficult to dig through such a boulder from down to up, so the worker prefers to do the other way around in order to have full control over the digging. In this way he can strike his pickaxe more strongly. Therefore he digs a diagonal tunnel from the main gallery roof toward the well, just over the boulder. This diagonal tunnel is named Bizesh which drains out water at the well, allowing the worker to dig the rest of well down to the gallery. As the worker digs down, the triangle between the gallery, well and Bizesh is removed gradually from top to bottom. That is why in such wells their bottom is much wider than their top, called "Shekam Darideh" (ripped belly) in local language (Fig. 4.24).

Fig. 4.24 Digging a Bizesh

4.16 Devil-Bizesh

Sometimes the Devil spot is misestimated and as a result the Devil Does not meet the well above, going some distance apart, or sometimes the Devil cannot be dug in the desired place due to a hard rock lying just there. In these cases the Devil is connected to the well above by a diagonal tunnel which can be regarded as a combination of Devil and Bizesh altogether. Usually Devil-Bizesh results from a mistake that the worker makes in designating the right spot for digging Devil. As soon as he notices he corrects his way by digging a diagonal tunnel toward the well, though he was supposed to dig a vertical Devil all the way but ended up in a Bizesh (Fig. 4.25).

Fig. 4.25 Devil-Bizesh

4.17 Dastak

Dastak is a short tunnel deriving from the side of main gallery. When the worker notices that the half-dug well has fallen apart from the main gallery by mistake, he manages to dig a short tunnel from the gallery until beneath the well and then digs a Devil upward. This solution is feasible only where the well is not far away from the gallery otherwise it is very costly as well as difficult to use for hauling the excavated soil. A Dastak rarely exceeds a few meters in length, and sometimes it is dug diagonally from the beginning to form a Bizesh reaching the well. In this case it is called Dastak-Bizesh (Fig. 4.26).

4.18 Pishvaz

Sometimes the workers dig a horizontal tunnel from the half-dug well toward the main gallery that lies at a lower level. Water seepage can no longer hamper the digging of tunnel as it goes above the water table. This tunnel is called "Pishvaz" which means "welcome" for it goes toward the main gallery to welcome it metaphorically. They connect Pishvaz to the main gallery by digging horizontally from Pishvaz end

Fig. 4.26 Digging a Dastak

to the gallery roof, where Pishvaz reaches the main gallery with a depth difference. Now they manage to dig and remove the soil from between Pishvaz and the main gallery, resulting in a gallery with higher ceiling but connected to that half-dug well in the end.

The workers usually proceed to dig a Pishvaz only when: (1) the well bottom ends up in a very hard formation whose extent is not clear, and the workers have to dig over it horizontally to find out where it ends. It is very difficult and costly to dig into such a hard formation from bottom to top, and the workers prefer to dig over it until they find somewhere suitable to get down. (2) the workers come across a lethal build-up of poisonous gas at the end of main gallery and they rush to connect the gallery to a well in order to ease the air circulation. (3) they are out of enough workers to carry the buckets of soil to an access well with a windlass above. In this case they hurry to reach a well nearby to be able to empty the gallery out of the excavated soil (Fig. 4.27).

4.19 Lateral Tunnels

As mentioned the most common type of lateral tunnels is called "Dastak" referred to as short tunnels deriving from the main gallery with a length of a few meters. Dastak is not only dug to build a "Bizesh" in order to reach a half-dug well from bottom, but it may also be dug for other purposes such as increasing the area of

Fig. 4.27 Digging a Pishvaz

water seepage into the gallery. In fact all the lateral tunnels are similar and they are nothing but side tunnels branching from the main gallery in different directions. What differentiates these tunnels from one another is their length according to which they have different names. For example in Taft, Yazd a tunnel whose length exceeds 50–60 m is no longer called "Dastak" but "Cheragh Sooz" that literally means "lamp killer".

A "Cheragh Sooz" lacks shaft wells all the way, and is dug just to increase water infiltration into the gallery in groundwater saturated area. Sometimes the end of the gallery after its last well is called Cheragh Sooz, though the word "Pishkar" is much more common. If a lateral tunnel enjoys some shaft wells, it is called "Shakheh" which may be kilometers long, like qanat of Qasabeh Gonabad which has numerous long branches. An estimated the mother well of this qanat is 300 m deep, which ranks as the deepest qanat in Iran and probably in the world. Qanat of Qasabeh Gonabad enjoys a main gallery and 7 side branches with 427 shaft wells altogether. Also the qanat of Firoozabad Majoomard in Yazd is an ancient qanat which enjoys numerous side branches. This qanat has three main galleries one of which runs across a barren land north of Dehnow along the road of Khavidak toward Salsabil district. Its second gallery is called "Zeynab Beygomi" which originates in Maryam Abad village and its third gallery called "Shakheye Shoor" comes from Zarandiyon.

These three galleries run across the lands of Salsabil and then join together and pass by respectively Jewish cemetery, Mahdi street, Kohneh Abshoor and Malmir districts and eventually enter the town of Zarch. At this point the qanat again divides into two branches one of which runs across the district of Sardeh and operates a watermill there, and the other supplies water to Khanghan district and turns another watermill.

Fig. 4.28 A typical Zorna which transfers a surface current to a gallery

4.20 Zorna

"Zorna" is a passage that directs qanat water to a less elevated level than expected. A Zorna may convey water to the earth surface, another shaft well or a qanat gallery. According to where Zorna takes water, it can be classified into three types which are all called Zorna in the qanat workers' terminology, as follows (Fig. 4.28).

4.21 Zorna – Earth Surface

This Zorna is built where the qanat exit is too far away from the cultivated lands. In other words if the qanat water appears a long distance upslope from the farmlands, the workers dig a well which ends up in another horizontal tunnel with a very gentle slope in order to convey water to the desired area. This additional well along with its horizontal tunnel is preferred to an open ditch because it does not expose water to evaporation and pollution. Thus water pours down the Zorna well which can be some meters deep, as soon as it comes out of the qanat gallery exit. The sides and floor of this well is shored up with stones all the way to prevent erosion caused by the falling water. Water pours down the well and then enters a horizontal tunnel which is similar to a typical qanat gallery at water transport section which gradually approaches the earth surface and intersects with it. This tunnel may have some other access wells, given its length, and sometimes water may flow down some successive

Fig. 4.29 Zorna – Earth surface

Zorna to reach the desired area. Qanat of Hasanabad Chil in Taft has a Zorna with a 5 km long tunnel (Fig. 4.29).

4.22 Zorna – Shaft Well

This type of Zorna is usually used to connect some adjacent qanats in order to form a network of qanats. This Zorna is like the previous type except it does not take water to the earth surface but to another well belonging to a qanat nearby. When qanat water appears on the surface, it pours down the Zorna well which is totally lined with stones against water erosion. Water travels through the tunnel which is connected to the Zorna well, until the tunnel opens up just on the side of a shaft well in another qanat. Thus water pours down this shaft well into another qanat and flows down its gallery. These two waters are mixed and then travel the qanat gallery together down to the exit point where they are separated by measuring the water flow, since the owners know how much their water is. Sometimes the two flows are not separated and used together to irrigate a unique farmland. For example there are three Zorna of this kind in the network of Ahmadabad qanats in Taft, Yazd.

In some regions in central plateau of Iran where qanats cannot provide enough discharge alone, they are connected by building Zorna in order to pool their waters and make up sort of qanat network. For example this phenomenon can be found in the valley of Taft-Farashah (Eslamiyeh) in Yazd, which is some 2 km wide. There are 12 qanats all running down this valley, intersecting at different points and directions forming a leaf vein pattern. In this valley no one can find a qanat that works alone, but their waters mix up with each other or even with the water of tube wells in order to create a greater flow and then are used together. This issue has something to do with the concentration of qanats within a relatively small area and is also associated with the low discharge of each qanat. So the qanats can provide more water if they join. In case the exits of two qanats are close and their related farmland lies nearby, their waters mix up in an open ditch on the surface, though most of the qanats in this region are connected through Zorna.

4.23 Zorna – Gallery

This Zorna has nothing to do with the networking of qanats, but it is a solution to a technical problem that some qanats may face. The well and tunnel of this Zorna are both underground and never open up somewhere. Actually this Zorna is built where the workers come across an impassible barrier while digging a qanat gallery, and they have to dig upward at right angle and then get horizontal again and dig forward just over the same hard barrier. This structure forms a Zorna though it is located in the middle of a qanat gallery totally underground (Fig. 4.30).

4.24 Intersecting Qanats at the Same Level; Shotor Galoo (Siphon)

In case two qanats intersect at the same level, a special structure named "Shotor galoo" is built to pass by the older qanat without digging into its gallery. Shotor galoo literally means "camel's throat", which is actually kind of siphon in which one end is higher than the other, allowing water to come up and flow into the tunnel on the other side due to this height difference. The workers burrow under the existing gallery in the shape of a siphon when they want to intersect another qanat while digging a new one (Fig. 4.31).

4.25 Intersecting Qanats at Different Levels

Sometimes it is inevitable to dig a qanat just over another one which runs in different direction. In this case the two qanats intersect but at different depths. If the depth difference between the two qanats is considerable, no special action would be taken for the amount of water percolating from the upper qanat into the lower one is almost negligible. In case the depth difference is less than 3 m, the workers have to do something to prevent water escaping from upper toward the lower qanat gallery.

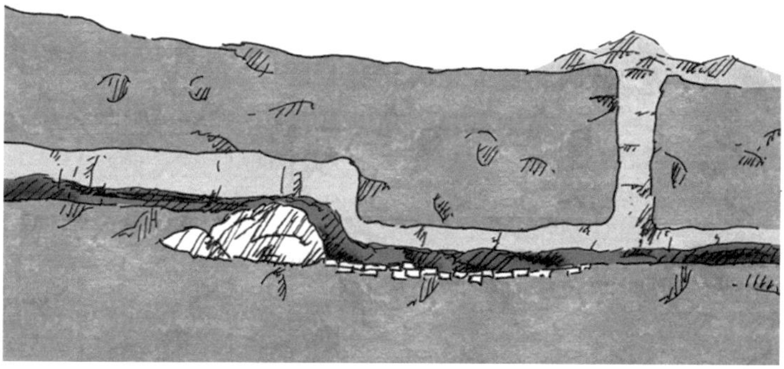

Fig. 4.30 Zorna – Gallery

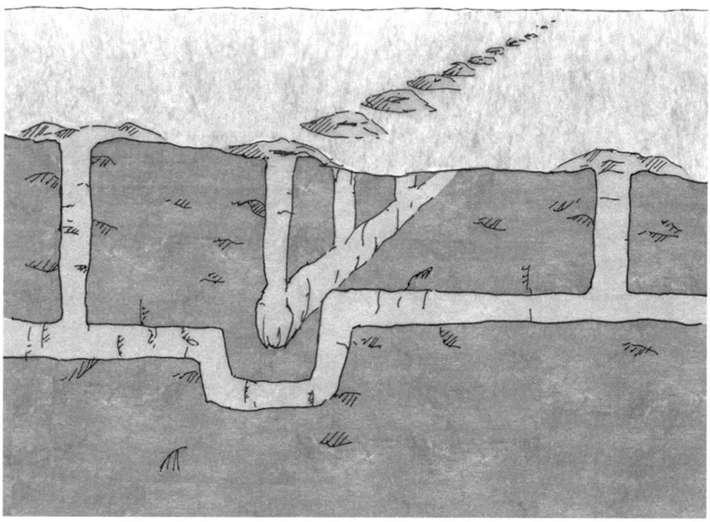

Fig. 4.31 Profile of a shotor galoo (Siphon)

At the point of intersection, they line the lower qanat gallery with ceramic or concrete hoops and at the same time they insulate the floor of the upper qanat with lime, cement, etc., of course with the permission of the owners of the older qanat. Doing so, water cannot escape from the upper qanat toward the lower qanat. Moreover the lower tunnel remains healthy as such percolation may cause collapse in its water transport section (Fig. 4.32).

Needless to say such an intersection is never allowed in water production section of qanat tunnel, and these practices are all applicable to water transport section. Before going around any kind of intersection, the workers should make sure that

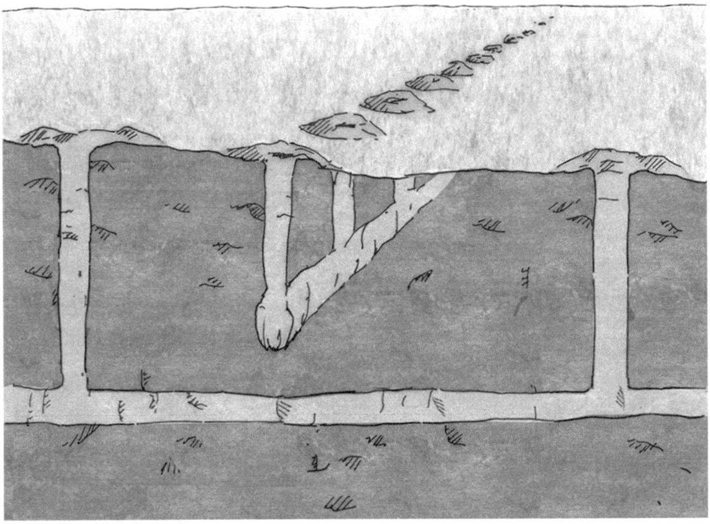

Fig. 4.32 Two qanats passing one another at different levels

both tunnels do not lie in water bearing zone. To do so, they dig a small hole into the side of tunnel and wait to see whether or not any water builds up in it. If not, they go ahead with the intersection. At the intersection point no shaft well is envisioned for the qanats, and their wells are dug separately some distance from the point. In some intersecting qanats, the workers devise an outlet through which the qanats can be connected when needed. For example if the upper qanat is not in good shape and needs repair, they divert its water to the other qanat for more convenience while working. No doubt connecting two qanats is contingent on the owners' consensus. The short tunnel or well which connects two intersecting qanats is called "Gozareh" which remains closed in normal situations. In case the lower qanat is much deeper than the one above, its gallery gets connected to the nearest well of the lower qanat by a short horizontal side tunnel which is called "Dastangineh" (Figs. 4.33 and 4.34) (Semsar Yazdi and Labbaf Khaneiki 2007).

Fig. 4.33 Gozareh connects two intersecting qanats

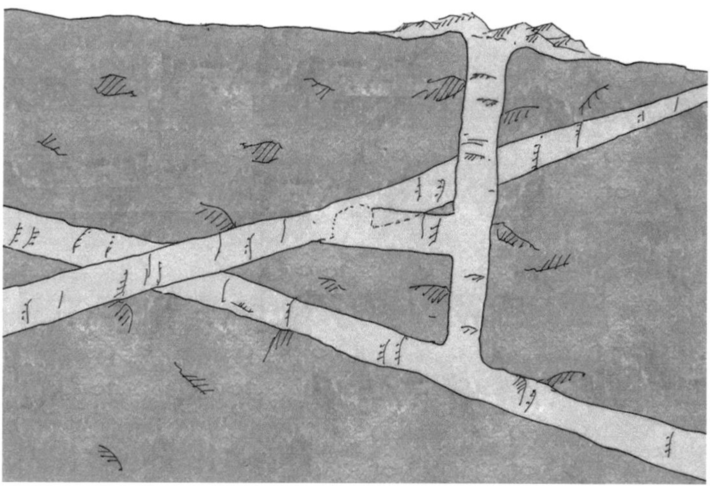

Fig. 4.34 A short tunnel which connects two intersecting qanats running at different levels (dastangineh)

Reference

Semsar Yazdi AA, Labbaf Khaneiki M (2007) Field studies on traditional qanat know-how and oral interviews with local practitioners. Yazd

Chapter 5
Barriers and Limitations of Qanat Construction and Rehabilitation

Abstract Qanat is always subject to a variety of barriers and limitations which may come about during its construction or refurbishment. This chapter classifies the possible barriers which may threaten the qanat practitioners or the whole project and then examines the measures the workers traditionally take to cope with barriers. The obstacles vary from poisonous gas to water build-up in the gallery.

Keywords Collapse • Safety • Reinforcing hoops • Poisonous gas • Water accumulation

5.1 Hurdles to Be Overcome in Construction and Rehabilitation of Qanats

Given that a typical qanat may be tens of kilometers long with several side branches spreading out in different directions, it is very likely for the workers to come across various geological formations while working in qanat. It is very difficult, costly and time consuming to dig through some geological formations, and sometimes it poses serious threats to the workers.

Over time the qanat masters have learnt how to deal with these barriers, building up kind of traditional knowledge which deserves to be mentioned here. In a nutshell the main potential problems that the workers may come across in qanat construction or rehabilitation are as follow:

1. Tunnel and well collapse
2. Hard rocky formations
3. Water build up
4. Underground swamp
5. Poisonous gas and vapor
6. Sediments
7. Vermin

A.A. Semsar Yazdi, M. Labbaf Khaneiki, *Qanat Knowledge*,
DOI 10.1007/978-94-024-0957-4_5

5.2 Tunnel and Well Collapse

The ground which is subject to collapse is called "Gonbate Zar" or "Shoolat" in Yazd local dialect. Such a soil enjoys instable and uneven particles which are not attached together firmly and easily caves in when a hollow space is dug out in it. Also some qanat masters believe that the shaft wells in which air circulation takes place from top to bottom are more subject to collapse. In this case the well begins to collapse mostly in the middle, forming the shape of a dome. That is why the workers call such wells "Gonbat" which means dome. In general it is very difficult and dangerous to dig a well in a crumbling soil, because the sides of the wells may collapse out of the blue and burry the worker inside. In case upper part of the well is more crumbling than its lower part, the lower part about 2 m to the gallery is dug in the shape of a funnel with sloped sides down to the gallery. This shape reduces the risk of sudden collapse. In the next sections, the traditional solutions to the problem of collapse would be examined (Fig. 5.1).

5.3 Protecting Qanat Workers Against Collapse

In crumbling wells an instrument named "Chapar" is used to protect the worker against the objects falling down, while working at the well bottom. Chapar is a semicircular wooden umbrella rotating around its stand and covering half of the well cross section all the time. Its stand is some 1.5 m high, which rests on the well bottom, providing a safe place against the soil and stones falling down (Fig. 5.2).

Fig. 5.1 Funnel shaped well

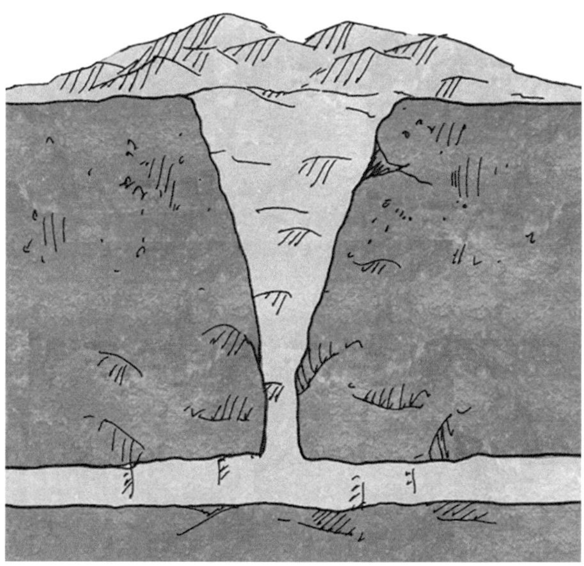

Fig. 5.2 Chapar

To better know how Chapar works, first we should remind how a well is dug. As mentioned a typical well has a circular cross section in which a worker sits and digs downward in a spiral way like a drill. Thus one side of the well bottom is about 30–40 cm higher than the other side, and the worker digs this way as if he is climbing down the stairs of a tower. The stand of the Chapar is placed on the higher side at the well bottom, and it turns around as the higher side shifts around the well. The worker moves the stand forward as he digs and turns around, so that half of the well is covered by Chapar to provide a shelter and the other half is left uncovered for the buckets of soil going up and down (Fig. 5.3).

5.4 Flood-Induced Collapses and Protecting Qanat Against Them

Some of the shaft wells may cave in totally in the wake of a flash flood rushing into it. Floods are among the most serious causes of qanat destruction, which can be curbed by some traditional methods. The workers pile the excavated materials around the well mouth to build a round dyke which can ward off the flood. These round dykes are called "Kamvar" and the distance between two "Kamvar" is called a "Poshteh". These "Kamvar" make up a landscape in an Iranian desert, which looks from air like a land overrun with rows of anthills (Fig. 5.4).

After the workers finish digging the well, they manage to line the well with bricks or concrete hoops though brick lining is more common. In a well such lining is usually done from its middle up. To do so, they cut a groove round the side of the well with a depth of about 15 cm, and then they put a row of bricks into the groove such that half of them stick out, providing a base for the lining. Afterward more bricks are stacked up and attached together with a mortar of cement or lime until it reaches the well opening. In the end a concrete round lid – called "Sargir" – is put

Fig. 5.3 Position of chapar stand at the *bottom* of well

Fig. 5.4 Kamvar, a row of kamvars and their view in the desert from air

on the well opening to block it. They may install such a lid in the middle of the well, which is called "Kamargir". Therefore there are three possible scenarios for shutting the well opening; (1) A round brick well may be built around the well opening on the ground surface up to 1–2 m high in order to stop the flood from entering the well, and then a round lid would be placed on the top of the aforesaid well. (2) The well opening may be covered by a round lid on the ground surface, which is called Sargir. (3) The well may be blocked in the middle by a lid called Kamargir (Fig. 5.5).

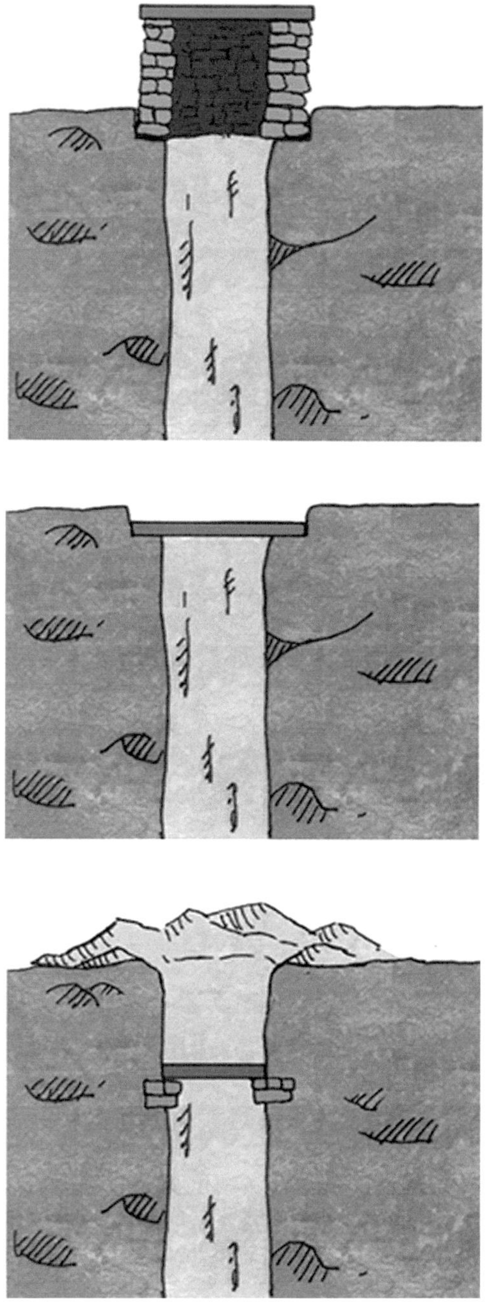

Fig. 5.5 Three possible scenarios for shutting the well opening

Nowadays the workers embed a row of bricks in the well side all the way round, when they want to build a base for a lid in the middle. In the past it was not usual to use bricks that way, but they install four long stones across from each other in the well side and then they put the lid on them. The long stones go 40–50 cm into the wall and wedged firmly in place with a piece of wood locally called "Gaz". They used to place one or two big flat stones on the aforesaid base, and then fill up the gaps and holes with sand and pebble and cover it with damp soft soil. In case flood is very likely in the region, the well bottom would be blocked too the same way. Nevertheless if a well collapses, it may be emptied though the workers prefer to dig a new well rather than toil away like that.

5.5 Strengthening the Crumbling Wells

A well that has collapsed is called "Pukeh". Such wells have lots of empty spaces in them, which may jeopardize the workers' lives if they do not know they are standing and working on a hollow space that may cave in again and make them fall down. To prevent such an accident a rope is tied around the worker's waist and the other end is tied to the stand of a windlass or a firm object nearby. The mud found in a well that has collapsed is called "Gele Morde" which means "dead mud". It is very difficult to empty such mud since new mud replaces and rushes back as the workers empty it. Recently the qanat workers have invented an interesting method to empty or even dig crumbling wells. They empty a small space at the top of the well in order to put a metal mold into the well. This mold is made up of five pieces which fit together to form a pentagon round the well. They observe a distance of 20 cm between the mold and the side of the well. Afterward they pour cement mortar into between the mold and well, burying six iron rods held vertically in there in order to strengthen the cement. After the cement solidifies, the worker climbs down the well through the mold and then digs the soil just under the cement so that the cement can slide down under the pressure of its own weight. The mold is placed on top of the previous cement and the same procedure is repeated, and then the worker digs underneath and pushes the cement down until it reaches the well bottom. Doing so, the workers can dig a well in crumbling soil without facing any collapse or having to haul any extra soil. It should be noted that the metal mold is not more than 80 cm in diameter.

In most cases where the workers do not encounter the so called dead mud, they use only earthen or concrete hoops to line the well, which is called "Towqeh". They put the hoops into the well one atop the other to shore up the well sides (Figs. 5.6 and 5.7).

Fig. 5.6 Metal mold used to line the crumbling well

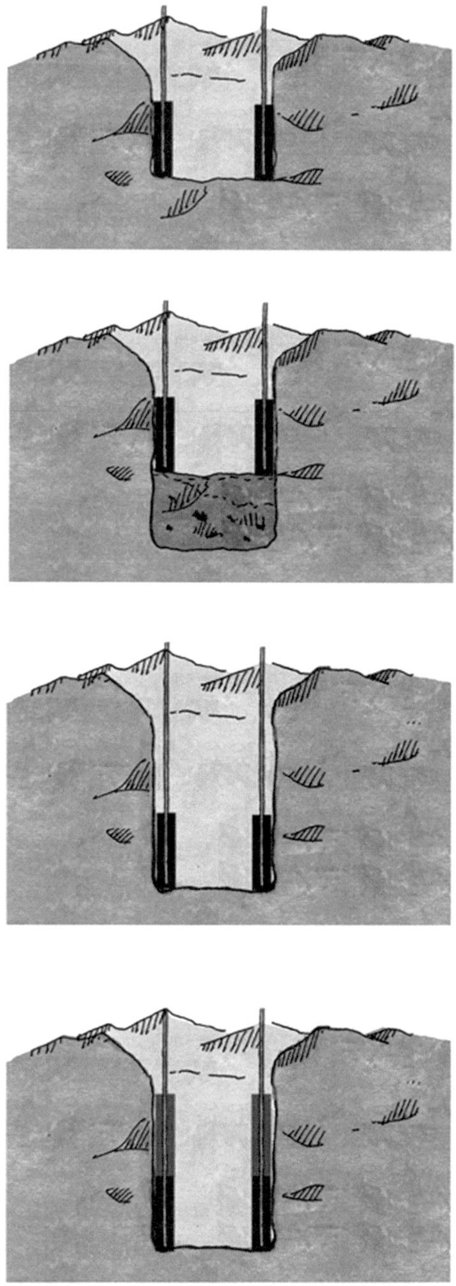

Fig. 5.7 Two types of tunnel lining with stones

5.6 Strengthening the Crumbling Tunnels

As for the qanat tunnel, the same "dead mud" can get the workers into trouble, so it demands a high skill to cope with it.

The worker should be able to recognize the dead mud and remove it all from the tunnel while digging or repairing the tunnel. If the worker is not skillful enough, he may treat the dead mud the same as he does in an intact normal formation. In other words he may dig through the debris which is disturbed and is no longer attached to the normal layers of earth. The mass of debris may cave in and burry the worker while he is digging a tunnel into it. Therefore a skillful worker differentiates a mass of debris from an intact soil and manages to remove the debris totally from the tunnel rather than dig through it even though this job results in an extraordinary large place in the tunnel.

The most common method to prevent a tunnel from collapsing is to use ceramic or concrete hoops (in Persian Kaval). In the mountainous areas where large flat stones or slates are abundant, the workers prefer to line the tunnel with stones rather than install hoops, because hoops are much more expensive than the natural stones strewn all over the area free of charge. Moreover hoops inevitably cover the tunnel floor as well and make it difficult to deepen the tunnel if necessary. Hoops are called by different named across Iran such as "Nay", "Kabar", "Gom" and "Kaval" out of which the word Kaval is more widespread. Kaval is oval in shape, formerly made of clay but now of concrete. Kaval should not be mistaken for the concrete rings used in shaft wells, because those rings called Towqeh are circular in shape with a diameter of 75–80 cm. Those rings are used to prevent collapse only in shaft wells, whereas shape of Kaval is such that it can resist more pressure on its top in the tunnel. The workers manage to install oval hoops as soon as the tunnel cuts through soft crumbling soils (Fig. 5.8).

In case exposure to air does not step up the collapsing of soil, the workers put off the lining until the tunnel reaches the next well; otherwise they install the hoops one

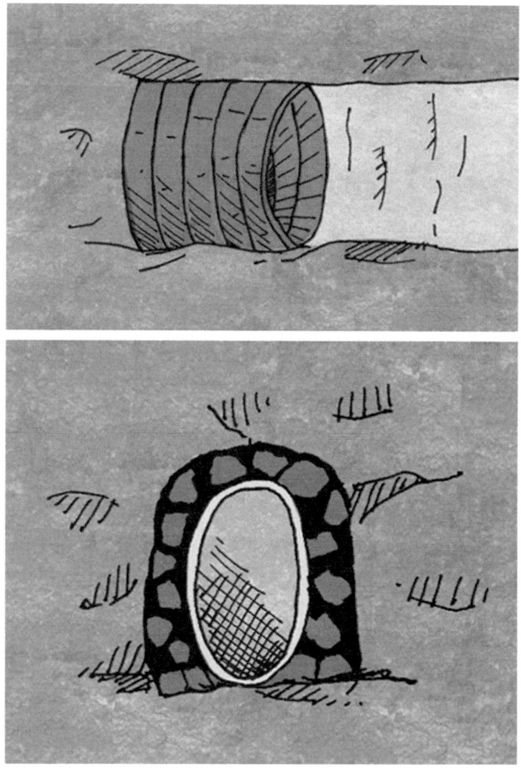

Fig. 5.8 Hoops (Kaval) used to line the tunnel

after another as they advance toward the next well. Sometimes the tunnel does not show any collapse until it reaches a well and air circulation takes place throughout the tunnel. In this case some pieces of clay and soil may detach from the sides of tunnel and fall down in the wake of air circulation. To prevent this, the workers use a type of Kaval which is specific to this kind of collapse. This Kaval has two parts with no floor, comprising of two curved pieces which form a horse shoe-like structure when fitted together. This type of Kaval is 140 cm high and easier to be carried along the tunnel, though it demands more skill to install and fix them in place (Fig. 5.9).

The qanat masters believe that it is not always possible to dig the tunnel first and then install the hoops. Some geological formations are so crumbling that they do not give the workers any opportunity to install the hoops after they finish digging the tunnel.

Sometimes the soil from the tunnel sides and roof comes down and suddenly collapses. When the tunnel cuts through such a soil, the digging of tunnel and putting the hoops should be done simultaneously otherwise a sudden collapse is very likely. This technique is called "Pish-Kaval" which means putting a hoop as soon as its place would be dug ready. In this method, the worker keeps a hoop on hand while

Fig. 5.9 Horse shoe-like Kaval

Fig. 5.10 A worker using the technique of Pish-Kaval

digging the end of the gallery just the size of a hoop, and he installs the hoop immediately after digging. Therefore he has to sit inside the hoops and install them one after another as he digs forward (Fig. 5.10).

The hoops used in such formations are oval in shape, integrated and smaller than the one before. This type is about 110 cm high and more difficult than the one before to be carried through the tunnel and inside the hoops to the end of gallery. The workers envisage a space of 4–5 cm between these hoops and the tunnel side, and the hoops are not allowed to be directly in contact with the wall and roof of tunnel, because some small lateral movements and pressures of soil may break the hoops if there would be no empty space. In the water production section, the hoops should not be fitted together completely with no distance between them, because it can insulate the tunnel and reduce the water infiltration. In this case, the workers consider a small distance between the hoops, and they put some pebbles and pottery pieces behind the hoops such that the distance would be covered not to allow the clay and mud to leak in, while keeping the water seepage into the tunnel. This practice is called "Palooneh". When the tunnel reaches a shaft well, the workers put four hoops across from each other just at the bottom of the well. They call such wells "Char-Kaval" which means a well with four hoops at its bottom (Figs. 5.11 and 5.12).

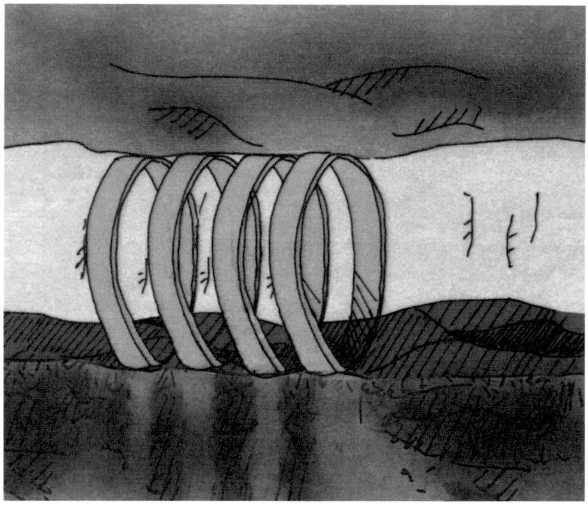

Fig. 5.11 Small gap between the hoops to allow water seepage

Fig. 5.12 Four hoops placed across from each other at the *bottom* of the well

The qanat masters believe that a skillful worker seldom gets trapped in a collapse while working, because soil lets him know before it comes down! A qanat worker should be fully sober and alert, paying attention to any indication that can be a clue to a sudden collapse. They recognize a soft clay formation in which the tunnel should be shored up with hoops otherwise a sudden collapse is inevitable. They call this formation "Gordeh-Gavi" which literally means "cow's flank", because pieces of mud the size of a cow's flank may fall down and destroy the tunnel. The workers say that a few seconds before the cave in, a sound like that of a handclap is heard. If they notice the sound, they would have the chance to escape. Also just before a collapse, some small pieces of mud start foaling down, indicating that a cave-in is around the corner.

5.7 Hard Rocky Formations

If the qanat workers come across hard rocky formations, they choose one of these options: they bypass it, burry it or dig directly through it. Where the tunnel hits a boulder, the workers widen the end of the gallery in order to better assess the extent and dimensions of the boulder. Doing so they realize which way the boulder is easier to be bypassed. They bypass the boulder and then get back to the general direction of the gallery, if the hard formation is not very extended. This practice is called "Baghal-Bor" (Fig. 5.13).

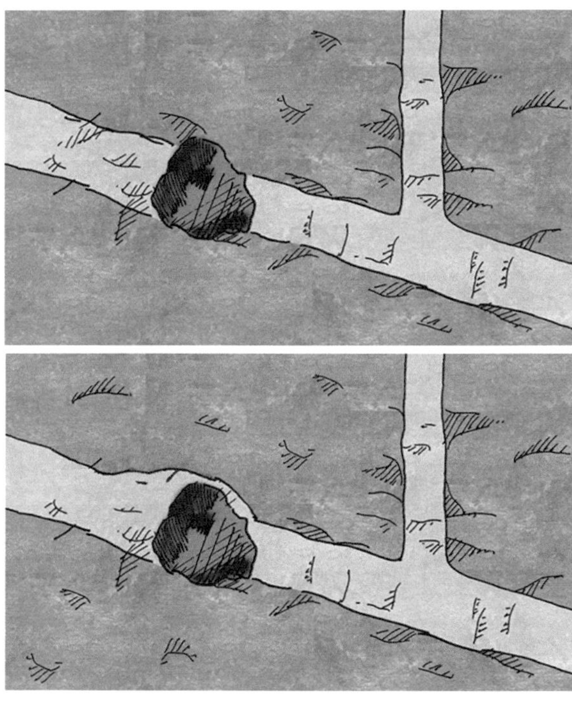

Fig. 5.13 Bypassing a boulder in the qanat gallery

Sometimes the workers encounter smaller boulders which can be moved or rolled. In this case they dig a hole just by the boulder and roller the boulder down the hole by means of a lever. The hole is dug on left or right of the boulder to bury it and get it out of their way. This hole is called "Goor" which means grave in Persian (Fig. 5.14).

Eventually if the workers come upon a layer of hard rock which is not possible to bypass or bury, they manage to dig through it inch by inch. In the past the only means to dig through such layers was chisel, lever and hammer which made this job extremely hard and laborious. During our field studies in Taft region, Yazd, we met up with an elderly qanat master who used to work in a qanat whose gallery cuts through a vast layer of sandstone. This qanat runs down the mountains of Shirkooh, in Nir area. The qanat has been called after its founder Haji Sadegh who died in early 1950s. This qanat was among the qanats which were owned by an individual rather than by a group of people like typical qanats. Haji Sadegh decided to extend the qanat gallery into the sandstone in the hope that the tunnel meets a rich aquifer beyond the hard layer. He remained steadfast in his belief that his qanat is going to get too much water after passing through the rocky layer, in the face of the qanat masters' objection. He urged the qanat masters to dig through the sandstone inch by inch, though he could not find the trace of water there. He spent all his money on this qanat until he went bankrupt, and resorted to selling out his farmlands for the sake of his qanat. It was ironic that he sold the lands which were supposed to be

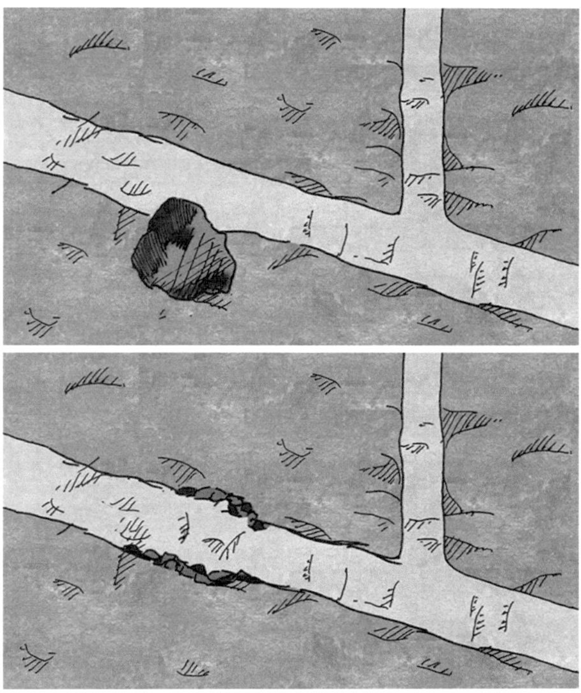

Fig. 5.14 Burying a boulder in the qanat gallery

irrigated by the same qanat if any water came out. The workers penetrated some 200 m into the sandstone and then gave up when it turned out that Haji Sadegh has died. This qanat is not only a good example of a rocky tunnel from technical point of view, but is also a manifestation of importance of qanat in the desert people's lives, and their iron will and determination.

Nowadays the workers use explosives as well to remove the hard obstacles on their way. Nowadays in Iran explosives for this purpose is in hand of ministry of agriculture, and the ministry sends the workers the required explosives along with an expert upon their request. In water transport section where the environment is dry both gunpowder and dynamite can be used, but in water production section dynamite is the only explosive applied. In dry parts, they drill a hole into the boulder and then put some gunpowder in it with two copper wires so close to each other embedded in the gunpowder. They seal the hole with a piece of tar and bring the other ends of wires onto the earth surface where they attach them to a battery. The electricity exchange between the two wires sparks and ignites the gunpowder. In case dynamite is not available, they thoroughly dry the hole and coat it with tar and then pour the gunpowder in it to blow up the boulder as mentioned. Recently technology has helped the qanat workers by introducing mechanical drills that can dig the hard soils more easily. Some of these drills work with electricity and some with pressured air produced by compressors.

5.8 Water Build Up

In the regions which enjoy rich aquifers, a large amount of water may be released into the gallery as it enters the water saturated zone. The workers have to continue their work and advance into the aquifer to a relatively large extent otherwise in some cases the current may dwindle. So the workers have to stay and work in water for many hours a day, which gives rise to a kind of skin disease locally called "Ab-Gaz" which literally means "stung by water".[1] Sometimes water build up is associated with an obstruction in the tunnel especially in the old qanats. Somewhere in the tunnel may collapse and block the water flow, leading to accumulation of water behind it. In this case water may rise and fill up the gallery of course up to water table not further. If the obstruction would not be found and removed, water comes up in the tunnel and soaks the soil and makes the tunnel cave in, especially where the soil is vulnerable to water build up. The obstruction should be removed as soon as possible, but first of all they should do something to control the water behind the obstruction (Fig. 5.15).

We interviewed a qanat master in Yazd, who remembered an accident that he had luckily survived in his youth. He dug through the obstruction directly from downstream upward, where the obstruction gave way out of the blue and fury of water

[1] This disease is known by skin rash and irritation which can be cured by an ointment made of pomegranate skin soaked in water.

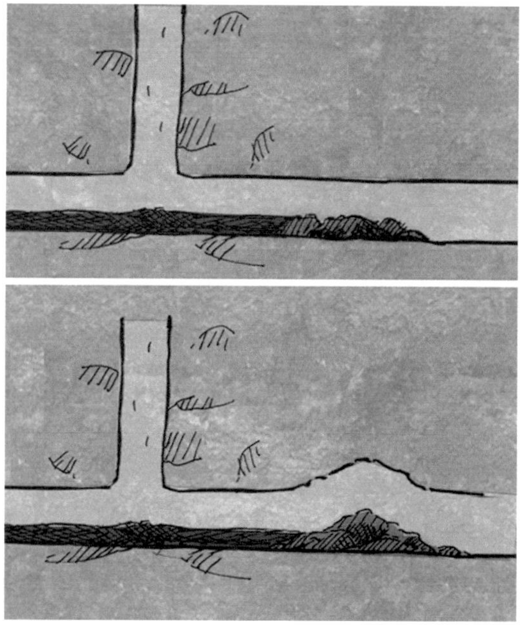

Fig. 5.15 Water build up in the qanat gallery due to the collapse of tunnel ceiling

grabbed him and took him as far as 200 m downstream. To prevent such accidents, the workers prefer to bypass the obstruction rather than dig through the blockage and release a large amount of water at once. This bypass is dug in a curved direction starting from near the blockage downstream and meeting the main tunnel some distance from behind the blockage upstream where the water can be harnessed. If the workers have to dig a long bypass, they prefer a rectangular one in order to avoid missing the right direction underground. In this case they dig just at right angle to the main tunnel and then turn 90° upslope and after a desired distance they again turn 90° toward the main gallery (Fig. 5.16).

Sometimes they envision some shaft wells for such a bypass to take the soil out and ventilate the tunnel, if the bypass is over 50 m long. Nevertheless they still fear that water rushes into the bypass as soon as it hits the main tunnel. To minimize this risk, the workers try to connect the bypass to the first well upslope from the obstruction. They tie some wood chips, straw and thorn into a bundle and push it down the well until it reaches the well bottom. They may thrust some other bundles like that into the well to make sure that water would not engulf the bypass when it opens up to the main gallery. Water would be filtered and calmed while passing through the bundles. They put some heavy stones on top of the bundles to keep them in the place under pressure until the bypass reaches the well. Sometimes they use some cotton sacks full of straw for this purpose. This way the water is emptied into the bypass gradually. Now the situation allows the workers go for the blockage to remove it if it is not too extensive otherwise they prefer to replace the bypass as the main gallery (Fig. 5.17).

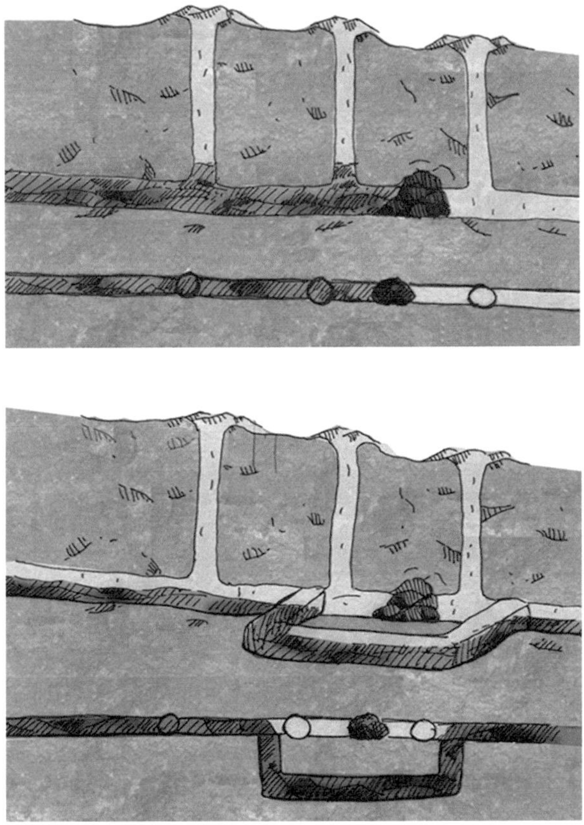

Fig. 5.16 Bypassing the tunnel obstruction in order to release water build up

Nowadays, it is usual to use the electric pumps to empty the well of water and make it possible for the workers to reach the blockage and remove it. Therefore the water can be lifted out onto the surface and the workers would be able to dig out the obstruction.

5.9 Underground Silt Pool

If the water transport section cuts through a soft sand formation, the floor of tunnel may be so saturated that it subsides. In this case the workers place some timbers of elaeagnus or willow wood on the tunnel floor so their feed do not plunge into the mud. Afterward they cover the timbers with slates and fill up the seams with soft clay to stop water from percolating downward. If the tunnel roof and sides are also subject to collapse, they have to install reinforcing hoops at the same time until they get out of the risky area. Sometimes the silt is so smooth and soft that the hoops sink

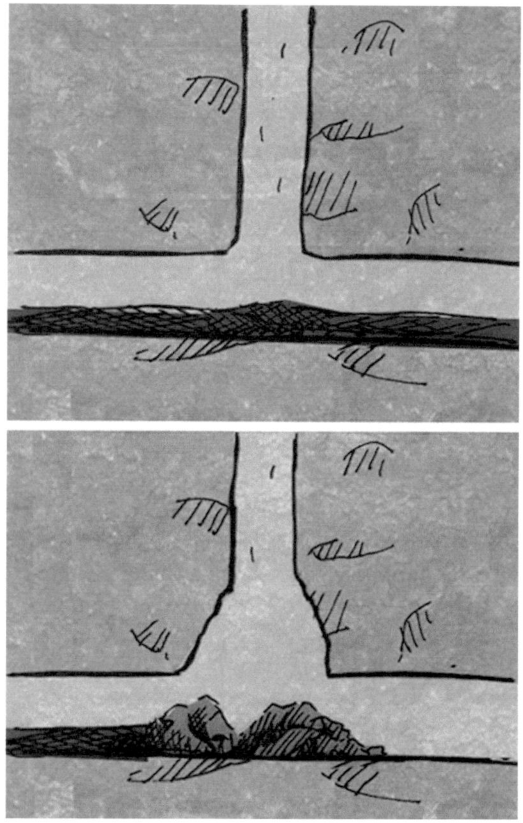

Fig. 5.17 Water build-up in a well and its releasing

into it when installed, so the workers have to build a base for the hoop first by dumping some stones on the tunnel floor and treading on them. They may bypass this troublesome area if no solution works.

Sometimes the workers may notice a crack or gap on the tunnel floor which drain out the water flow in water transport section. These cracks or gaps are called "Shaq" which should be filled up with lime and clay otherwise water escapes through them and the qanat discharge dramatically decreases. If the workers find a "Shaq" in water production section, they examine it to figure out whether it gives or drains water. If it drains they fill it up with kneaded thick clay mud otherwise they do not manipulate it.

5.10 Poisonous Gas and Vapor

The qanat workers use two separate words for the poisonous gas in qanat, each of which has a special connotation; "Gaz" or gas and "Dam" or oxygen shortage. According to them "Gaz" is continuously being re-produced inside the ground so that one can see its tiny bubbles coming up in water. They believe that Gaz is more common in mountainous qanats and becomes more plentiful during cloudy days. In Garizat and Bonadkook in Yazd there exist many qanats with harmful gas which have claimed some lives so far. Presence of harmful gas can pose a serious threat to the workers and even bring their work to a standstill. Sometimes the amount of gas is so much that it comes out of qanat exit and even suffocates the birds that descend to drink water. The water of such qanat tastes sour, which gets better as the water travels some distance on the ground. Gaz is persistent and remains in a particular part of the gallery as the soil gives it off. According to the qanat workers, it is hard for them to breathe in Gaz though their lamp does not go out. In this case they have to pour damp sand down a well nearby in order to push up the stagnant air and make a circulation in the tunnel. Sometimes they mix the sand with vinegar to better fight the Gaz. They may have to repeat this job once in a while before being stifled by the Gaz. Some workers make the air circulate by lighting a fire in the tunnel just at the bottom of a well, because warm air starts ascending and fresh cold air goes down from the top of well to replace it.

In addition to these methods, the workers may resort to some other ways to tackle the problem of gas. For example they may manage to dig a twin well as mentioned before. A twin well can ease the air circulation which can be stepped up by pouring soft sand down a well to propel the air up the other well (Fig. 5.18).

Sometimes the workers tie some woods and thorns into a bundle and set fire to it and then drop it down a well where the workers are in trouble with gas. The flaming bundle is dropped down the well to move the stagnant air up. They have to do so once every 2 or 3 h to disperse the gas and relieve the workers of the stuffy air.

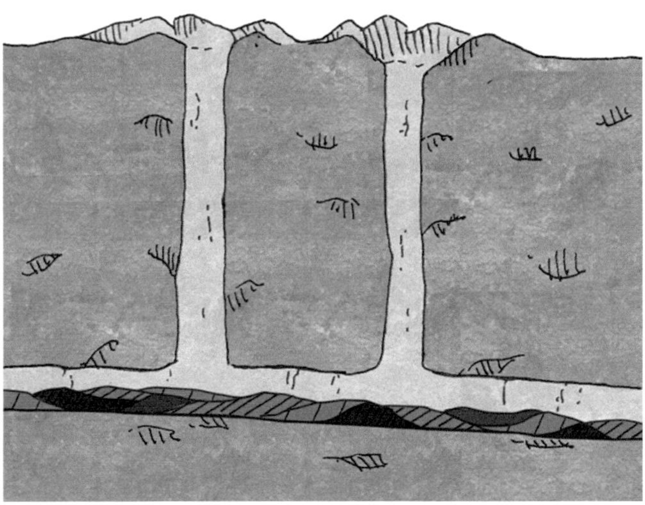

Fig. 5.18 Twin well for ventilating the shaft well

"Dam" implies oxygen shortage which takes place in the wake of air stagnancy, for the oxygen molecules would be absorbed by the mineral materials in the closed space. Dam is easier than the "Gaz" to fight off, because it is caused by air stagnancy which abates after a while. According to the qanat masters, "Dam" puts out a light or a candle where it builds up in the tunnel. Before the workers go down a well, they tie a light to a rope and send it down the well to see if the light keeps burning or goes out to make sure that there is enough oxygen down there. The solution of twin wells is also applicable to the problem of "Dam", though there are other methods practiced in this case as well. One of these methods is to use slaked lime to absorb as much carbon dioxide as possible in the tunnel. First they mix some quicklime with water to hydrate it as Ca(OH)2 which can react with carbon dioxide to form calcium carbonate.[2]

Some workers believe that keeping a bowl full of vinegar beside them or some cloves of garlic can be helpful, while they are working in a stuffy tunnel. Finally the most common method to curb the poisonous gas is to build a ventilating duct locally named "Havakesh" which literally means chimney. This duct can be built only in the tunnels which have been lined with ceramic or concrete hoops, because the 20 cm space between the tunnel roof and the hoops serves to move the stagnant air in the tunnel. To better understand how this ventilation system functions, we name the last well just before the gallery end "well A" and the well down slope from it "well B". Imagine the tunnel from well B to well A and then farther to the gallery end has a lining of hoops all the way. Now the workers cover the opening of hoops with a thick canvas curtain at the location of well B in order to stop air from entering the hoops. Therefore the air coming down the well B has no way but to enter the space left between the tunnel roof and hoops, which leads to the well A. At the bottom of the well A, the ducts above the hoops on both sides of the well are bridged with a canvas hose which transfers the fresh air from one side to the other side of the well A, and the air continues its way on top of hoops to the gallery end where the workers need it to breathe. The air gets back to the well A through inside the hoops and then goes up the same well, forming a constant circulation of air. The mentioned hose is made of canvas with some rings made of pomegranate wood inside to keep it open. The hose is fixed on the wall in a curved way to prevent any interference with the work of windlass and the buckets going up and down the well. This technique allows the workers to have fresh air as they advance into the earth. They install new hoops as they dig forward in order to extend the duct that can bring fresh air. Sometimes one can find a gallery lined with hoops, where there is no trace of collapse or crumbling formations. If we remember the ducts above the hoops and the air conveyed by them, we realize that it still makes sense to install hoops in a stuffy tunnel though it is not likely to collapse (Fig. 5.19).

Some workers prefer to use other techniques to ventilate the tunnel rather than resort to hoops. They put a bellows on the ground surface just by the well mouth and pump the fresh air into a canvas hose attached to the bellows. The other end of the hose comes to the gallery end where the workers are busy digging. In a mild condition,

---

[2] According to this reaction: $Ca(OH)_2 + CO_2 \rightarrow CaCO_3 + H_2O$

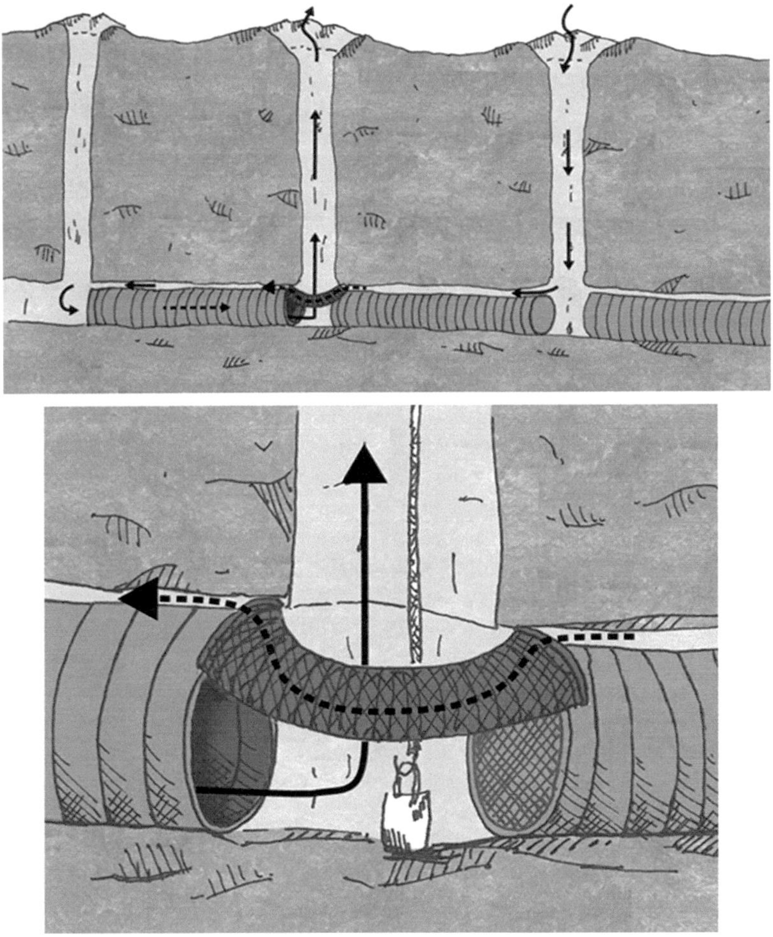

Fig. 5.19 Ventilation of qanat tunnel by using reinforcing hoops (havakesh)

a worker may move quickly along the tunnel to ease the air circulation, though nowadays electric fans are being used widely as a modern substitute for the ancient ventilation techniques.

According to the qanat masters, the dimensions and shape of a gallery can play an important role in ventilation. The gallery should be 60 cm wide all the way where the workers are in trouble with gas. They use a special word “Araj” which means a tunnel with a width of 60 cm all the way with no bulge or dent on the walls. They believe any slight unevenness on the walls of tunnel can hamper air circulation, resulting in stuffiness of the tunnel. Therefore they dig the tunnel in such a situation more carefully.

5.11 Sediments

According to the traditional qanat masters, the faster the water flow in the tunnel, the more sediment would be settled. They believe that a relatively steep tunnel in mountainous areas more stirs the water leaving more sediment on the floor and sides of the tunnel. It is worth noting that this issue has a scientific basis, because the water with soluble calcium bicarbonate can leave calcium carbonate, water and carbon dioxide where it flows down at a relatively high speed.[3] This calcium carbonate accordingly settles on the tunnel floor and forms what the qanat workers point out. That is why those mountainous qanats with calcium bicarbonate enjoy fresher water with less soluble substances at the cost of massive sediments especially in the water transport section, which can clog up this part of tunnel if not removed. This sediment is locally called "Shæ" which should be removed and lifted out every once in a while, otherwise the qanat tunnel gets clogged up or becomes "Koor" (blind) as qanat workers say. In the mountainous qanats the sediments are denser, harder and more difficult to be broken and removed. The workers do not scrape the sediments off the tunnel walls totally but they try to leave a thin layer of them to keep water percolation at bay in water production section.

5.12 Vermin

Such dangerous animals as snake are likely to show up in the qanat wells and tunnel, bringing the work to a standstill temporarily. Snakes found in a qanat are those that have fallen down a well while foraging for food or looking for shelter. These ectotherm creatures lose their body's temperature when they fall into the qanat water and then go numb so that they cannot be as nimble as they are on the ground surface. Therefore the worker uses a Y shape stick to catch the snake and put it in a cotton sack and send it up along with the spoil in the bucket. Some other animals like fox may have fallen down the qanat wells. They never hurt them but they catch them and lift them out alive.

According to qanat masters, porcupines like dry abandoned qanat galleries to make their dens there. If people want to rehabilitate dry tunnels they should watch out for possible porcupines that may panic and hurt the workers with their spines. Also bats are among the animals that qanat workers may encounter in the qanats. These bats are small in size and insectivore (Semsar Yazdi and Labbaf Khaneiki 2008).

Reference

Semsar Yazdi AA, Labbaf Khaneiki M (2008) Field studies on traditional qanat know-how and oral interviews with local practitioners. Yazd

[3] $Ca(HCO_3)_2 = CaCO_3 + CO_2 + H_2O$

Chapter 6
Tools and Equipment

Abstract The qanat workers have many tools and equipment to dig the well and tunnel, haul the excavated soil on to the surface, measure slope and height difference, illuminate the work place underground and ensure their own safety. They have also invented amazing tools to calculate time of irrigation and gauge water flow. In this chapter we show how elaborate their tools are in order to work out qanat construction and maintenance under every circumstance.

Keywords Digging tools • Illumination tools • Measuring tools • Safety tools • Workers' accommodation

6.1 Wooden Windlass

Windlass is kind of pulley by which workers can lift out the spoil from a qanat well. A windlass is composed of four timbers crossing each other two by two, and four poles go through the crosses at right angle to attach them together and form a pulley. The two crosses are placed parallel about 1 m away and then nailed and tied to the four poles to get them firmly attached. A stronger pole goes through the middle of the pulley to act as an axis around which the pulley rotates. This axis is placed on two wooden stands which are firmly fixed into the ground just on two opposite sides of the well opening. A noose is tied around one stand which is placed around one end of those crosses to brake the windlass and keep it motionless when needed. This noose is called "Pavandeh", and the windlass resumes working if "Pavandeh" releases it (Fig. 6.1).

To install the windlass, first a worker stands by the well and stretches out one hand toward the middle of the well and then releases a stone at that point. If the stone falls down right at the middle of the well bottom, the worker stretches his both hands opposite each other while holding a stone in each hand. The worker releases the stones and watches where on the ground they land. He installs the two stands of his windlass wherever the stones hit the ground. It is very important to place the windlass just in the middle of the well otherwise the bucket would collide with the side of well and rupture over time (Fig. 6.2).

A.A. Semsar Yazdi, M. Labbaf Khaneiki, *Qanat Knowledge*,
DOI 10.1007/978-94-024-0957-4_6

Fig. 6.1 Wooden windlass

Fig. 6.2 Worker determining the place of the windlass stands

According to how the rope is tied to the windlass, there are two types of windlass. The first type is named "Yek-Sar" or simple windlass which works only with one rope and one bucket hanging from it. The second type is named "Do-Sar" or double bucket windlass which enjoys two ropes with two buckets working parallel. In deep wells the workers prefer to use the second type in which one bucket goes down the well while the other is coming up. Thus they economize on time and energy (Fig. 6.3).

The weight of the empty bucket going down helps the worker pull up the full bucket coming up. The two alternative buckets can also accelerate the work. In some wells the weight of the empty bucket and its rope is so propelling that the workers need to control the speed of windlass rather than push it turn where the full bucket reaches near the well mouth. The traditional qanat workers usually use two

Fig. 6.3 Double bucket windlass

different types of rope. The first is made of date palm fibers, called "Sazoo" which is used in shallow wells and on simple windlasses. Privilege of this rope is that it would not be soaked in water and its weight does not change. The second type is made of cotton which is suitable for the double windlasses.

As mentioned, when the bucket reaches the well top, the windlass should be braked by means of a noose that is put around one of the arms of the windlass to harness it, otherwise the rope suddenly unrolls and hurts the worker (Fig. 6.4).

In the past the qanat workers used to measure the depth of the wells by the length of the rope sent down. For example if the poles of a windlass are 1 m away from one another, each turn of the windlass unrolls 4 m of the rope, because the windlass has four poles. If the windlass rotates two and half times, it means the rope has gone ten meters down the well because: $2.5 \times 4 \times 1 = 10$ (Fig. 6.5).

Sometimes the workers use serial windlasses in a well, which is a very painstaking job. In very deep wells where the weight of long ropes does not allow the windlass to work properly, every 80 m in the well a small chamber is dug into the ground in order to install additional windlasses. Each chamber is a station in which there is a man operating a windlass which receives the buckets from the lower station and delivers them to the upper one. Henri Goblot describes this technique in his book when it comes to the qanat of Qasabeh Gonabad. "… They did not give up when they did not find water at the depth of 100 m but they built a chamber at the bottom of their well and installed another windlass there and then went on digging. In fact at that depth they dug another well but not exactly in the same direction as that of the previous one. The second well was dug with the depth of 100 m just a few meters away from the first one, though no trace of water was found there. Therefore they resorted to digging the third well which eventually reached water at the depth of 100 m. The spoil was hauled from here to the second well and then to the first

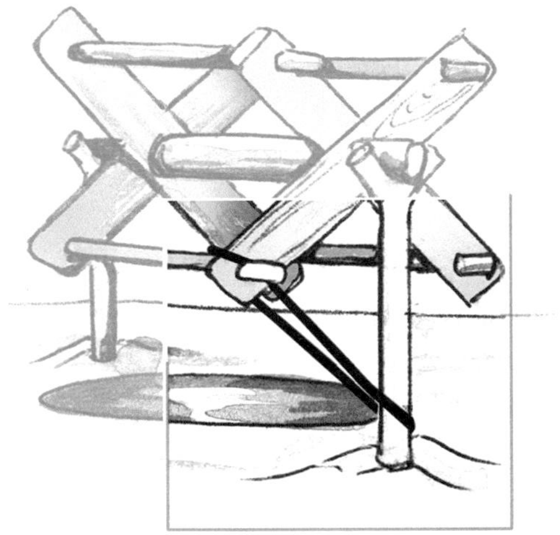

Fig. 6.4 Windlass brake

Fig. 6.5 A typical windlass

well by means of the windlasses installed at the bottom of each well. Doing so they could overcome the technical barrier in using one single windlass to dig a 300 m deep well like that…" (Papoli Yazdi et al. 2000: 122–127) (Fig. 6.6).

Another hypothesis is that the three wells were dug as an integrated well at once at such a depth, but stations to collect the spoil were devised every 100 m in the well (Fig. 6.7).

6.2 Rope and Types of Knot

In qanat two types of rope are common the most; cotton rope locally called "Arghamchi" and rope made of date palm fibers named "Sazoo". Arghamchi has strong fibers, which is made of three or four smaller cotton ropes interwoven. This rope is used mostly in double windlasses which need stronger and thicker ropes,

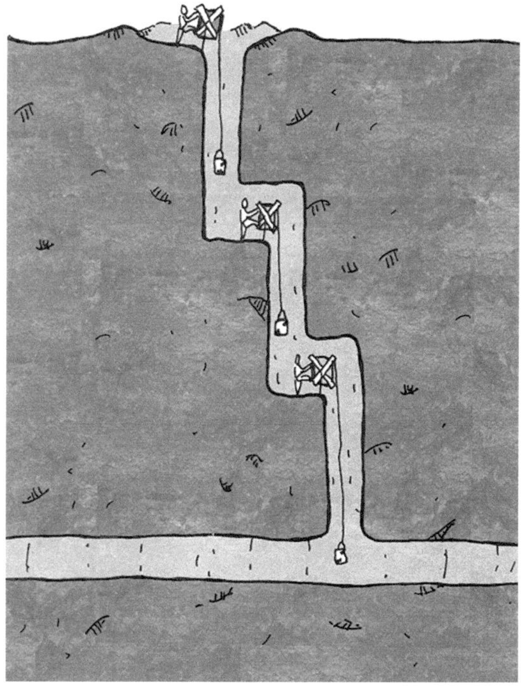

Fig. 6.6 Serial step-wise wells

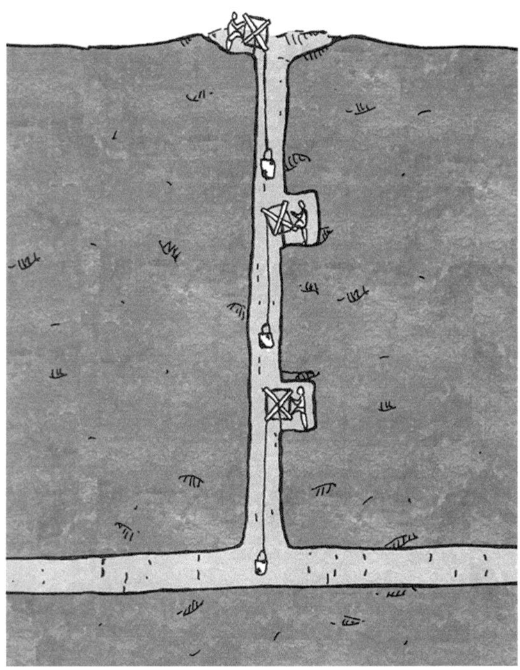

Fig. 6.7 Serial windlasses in a qanat deep well

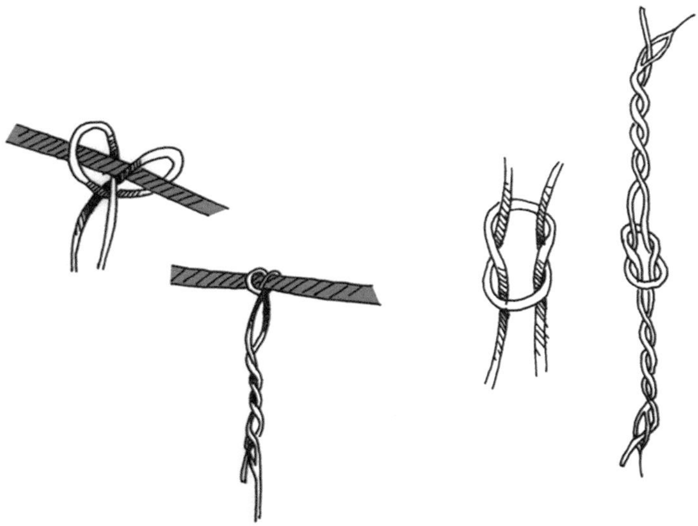

Fig. 6.8 Dogali knot

though its disadvantage is that it absorbs water in wet conditions and gains weight over time. Moreover this type of rope comes to rot and fall apart when wetted for a long time. Therefore the qanat workers prefer to use the other type of rope Sazoo in water production section, because its fibers absorb much less water. To make Sazoo, the date palm fibers are woven together to form some thin strings which are called "Fetileh" and then each two strings are twined to form a primary rope which is called "Menal". Afterward they use a special weaving machine to twine the Menal and turn them into Sazoo. Sazoo is stronger than the cotton ropes, though it may decay too if it remains in contact with water for a long time.

When a rope is not long enough, the workers tie it to another rope with a special knot named "Dogali". They twist the two ropes and knot them such that they attach more strongly as they are pulled apart. The workers say that this knot is fully reliable when they want to send people down a well (Fig. 6.8).

Nowadays using wire ropes is catching on in the country, replacing the traditional ropes. The workers have to smear their wire ropes with grease to prevent them from rusting in wet environments.

6.3 Illumination Tools

In the past it was common to use a ceramic pear-shaped lamp with a small handle at the back. This lamp was fueled with vegetable oils such as oil of cotton seed, castor or sunflower. The lamp had a spout on which a wick was placed with its other end soaked in the oil which went up the wick and burned with a yellow flame. The workers always had on hand a container full of oil to re-fuel the lamp whenever needed.

They built a niche on the gallery wall to place their lamp there and they also devised a cover for the lamp if water dripped on it. In the old qanats one can still find the black marks of those ancient lamps on the tunnel walls and roof (Fig. 6.9).

Later another type of lamp was introduced, which worked with petroleum, locally called "Cheragh Mushi". Disadvantage of this lamp outweighed its advantage, because it produced more smoke and made it difficult for the workers to breathe after a while in the tunnel. That is why this lamp was put aside soon and was replaced by a much better alternative carbide lamp or "Cheragh Karbit". This lamp was invented in 1892 and then introduced and widely used by qanat workers across the country. Even today this lamp could have retained its position despite availability of electric torch (Fig. 6.10).

Carbide lamp is based on a chemical reaction between calcium carbide (CaC_2) and water which produces acetylene (C_2H_2). This lamp consists of two small containers one on top of the other. The upper container is filled with water which drips out of a tiny hole at the bottom. Some calcium carbide is put in the lower container

Fig. 6.9 Oil lamp

Fig. 6.10 Petroleum lamp (Cheragh Mushi)

Fig. 6.11 Carbide lamp

where it reacts with the water and produces acetylene gas. This gas gets out through a small pipe attached to the lower container, and burns with a bright white flame. When all of the carbide has been reacted, in the lower container a wet paste of slaked lime (calcium hydroxide) is left. 200 g of calcium carbide can ensure 8 h of illumination for a typical carbide lamp. Many qanat workers prefer this unfocused light to that of electric torch. This lamp does not leave smoke and fume like the other types do. In some situations the qanat workers cannot use any kind of illuminating tools even carbide lamps. For example when they manage to dig a "Devil", they work completely in the dark. They dig overhead from the tunnel to a well above in the dark without losing their way underground (Fig. 6.11).

In general carbide lamp enjoys three advantages as follows: (1) it can work continuously and for a long time (2) its light is considerable (3) it does not give off fume. Therefore carbide lamp overdoes the former technique for illumination such as oil lamp or petroleum lamp.

6.4 Pickaxe

To dig qanat tunnels and wells, according to their function in general two types of pickaxe are common to be used as: well digging pickaxe, tunnel digging pickaxe. Well digging pickaxe has a relatively short handle which is some 30 cm long and weighs 2–3 kg. This pickaxe is short handled because it is used for digging wells which are not more than 80 cm in diameter and a longer handle is not easy to wield there.

Tunnel digging pickaxes fall into 5 types based on their function, each of which is used in a particular part of tunnel cross section. For digging the tunnel ceiling, a type of pickaxe named "Boone kani" is used, which weighs 1 kg to be easily handled while digging overhead. Afterward the top of tunnel cross section is dug by the pickaxe of "Karkani" that weighs 1–1.5 kg. The lower part of the cross section -between the top part and the middle part- is excavated by "Pakar Kani" pickaxe that weighs 2.5 kg.

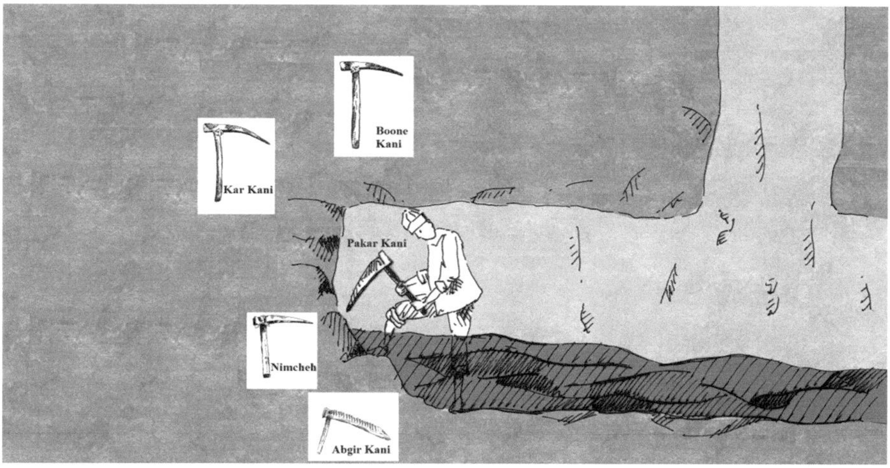

Fig. 6.12 Types of pickaxe used in different parts of tunnel cross section

When the workers want to dig the upper part of qanat tunnel in order to extend it, they prefer to use these types of pickaxe which are light enough and do not tire them out. Extending a qanat tunnel is a time consuming job which is done by means of different types of pickaxe according to the situation of tunnel. Now from the middle of the cross section to its bottom, two other parts are considered out of which the upper one is dug by "Nimcheh" pickaxe that weighs 3–4 kg where water current is 20–30 cm deep, and the lower part is dug by "Abgirkani" pickaxe where the depth of water reaches some 80 cm. Abgirkani pickaxe is the longest pickaxe with an 80–90 cm long handle, weighing between 6 and 9 k. This pickaxe is used for digging in water production section where the water is over knee high and a long handled pickaxe is more suitable to dig under water (Fig. 6.12).

Apart from the aforementioned types of pickaxe classified based on their function, one may see more types which can be classified mostly based on their shape as follows:

- Nok Gonjeshki, sparrow's peak like pickaxe for digging into rocks
- Barg-e Bidi, willow leaf like pickaxe for digging in soft soils
- Dandane dar, ragged pickaxe for digging in mud
- Sang Kani, stone digging pickaxe for hard formations
- Saghf Kari, roof building pickaxe for digging the tunnel roof in the shape of a pitched roof
- Tabari, hatchet like pickaxe which has a short and flat end for digging hard soils
- Hammer and chisel for digging rocks and very hard soils

Fig. 6.13 Forging devices

6.5 Forging Devices

The pickaxe would be blunted after a while, and the worker has to send it up in order to sharpen it again. To do so, the worker needs to have the necessary tools around the well opening. To sharpen a pickaxe, three people are needed to team up at the same time; a worker should operate a bellows to fan the fire, a worker should hold the pickaxe on the forge with a pair of pincers, and a worker hammers the heated pickaxe such that its end would be sharpened again. It should be noted that this job demands a high level of skill and expertise. When the pickaxe would wear out over time, they make a cut at its end and embed a piece of iron in there and then weld them together. This repair is usually done by a local blacksmith, because it goes beyond the qanat working profession (Fig. 6.13).

6.6 Shovel

This type of shovel is smaller than a normal one, with a shorter handle which can move about in such a small place underground. This shovel is used to gather the spoil and dump it in the buckets in order to lift it out. A typical qanat shovel enjoys a 15 cm handle in length (Fig. 6.14).

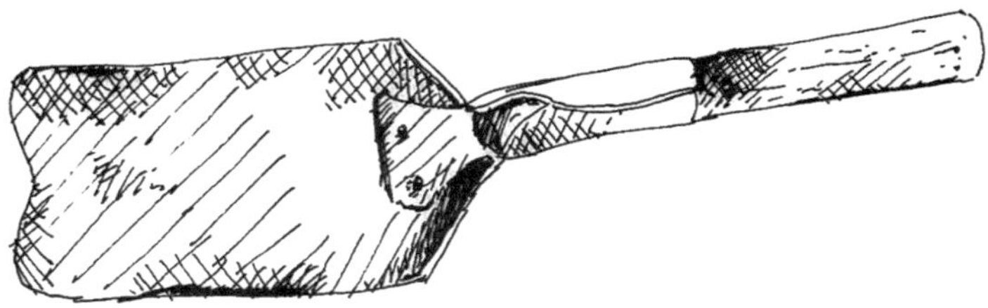

Fig. 6.14 A typical qanat shovel

6.7 Bucket

Dalve, Bucket is used to lift out the spoil and haul it onto the earth surface. In the past the buckets were made of tanned sheep skin, locally called Timaj, but today rubber is used to make buckets called Manjil or Manjidi. In the past the workers sewed a tanned sheep skin round with some leather strings cut out of the same skin. Afterward the rim of the bucket was folded back and sewed so that a flexible thin branch of pomegranate can drive into it all the way round. This way the bucket remained open, ready for use.

Moreover it was necessary to devise some methods to better maintain the buckets which were subject to getting rotten or worn out. Therefore the workers used to bury their buckets under lime in order to prevent them from rotting and decaying. In the past the buckets were sometimes made of cow leather. Also kind of container was used to carry soil, named Zanbil which was made of date fibers. Zanbil was not suitable to carry mud and wet soil because it was torn up easily. The size of the bucket depends on the amount of spoil, type of windlass and the windlass worker's strength. In the mother well where the worker has to send out a longer rope, smaller bucket is preferred because the rope is so heavy there and a heavy bucket may make it impossible for the worker to pull them together up. In contrast, in the shallow wells the workers prefer using big buckets (Fig. 6.15).

At present modern technology helps overcome some limitations of traditional methods, like using electric lifts in order to pull up bigger buckets even in deep wells, whereas in traditional method it is the number and strength of workers which dictate the size of bucket. In the very deep wells the wooden windlass may buckle under the pressure of the long rope hanging in the well. For example in the qanat of Gonabad, east of Iran, lifting out a bucket of spoil in its deepest well entails a 300 m rope which weighs at least 150 k. This rope is three times as heavy as a normal bucket full of soil hanging from it, so a traditional windlass cannot stand under such a weight. As mentioned, to solve this problem the workers build a windlass station every 100 m in the deep well, so there were three windlass working together to pass on a bucket from one to another. Each station was a chamber dug along the well so that a windlass could be put up there and a worker could work with it. The first windless pulled up the bucket from a depth of 100 m and the same bucket was hooked to another rope overhead pulled by the next windlass. The third windlass was on the ground surface (Papoli Yazdi et al. 2000: 133).

Fig. 6.15 Bucket with different sizes

6.8 Chapar, Wooden Shield

Chapar is a semi circle wooden shield with a wooden stand which is placed on the floor at the well bottom. The worker takes shelter under the shield to protect himself against the falling objects. Given that a bucket goes up and down the well all the time and it always collides with the sides of well, stones may fall off the well and hit the worker. Also the bucket may detach from the rope and fall down or the workers on the ground may knock something down the well unintentionally. All of these are fatal to the worker at the well bottom. The shield is such that half the well is always covered and half is open to allow the bucket in and out. The shield turns around its stand which is 1–1.5 m long as the worker digs the well in a spiral way (Fig. 6.16).

6.9 Tool for Calculating Times

To measure the time every shareholder has for irrigation, the qanat beneficiaries have invented a special type of water clock or clepsydra. Their clepsydra consists of two bowls made of copper one of which is so small that could freely float on the surface of water in the large one. The floating bowl has a tiny hole at its bottom through which water can enter the bowl and gradually fill it up. After being filled which may take a certain time, the small bowl sinks in the water and bumps into the bottom of the large bowl. As soon as the bump would be heard, a unit of time would be over, so the time between the two bumps equals a certain unit of time. One can

Fig. 6.16 Chapar to protect the workers

also find some marks cut into the inner side of the small bowl which divide the certain unit of time into the shorter fragments. The time it may take the small bowl to be filled and sink varies from area to area in the central plateau of Iran (Fig. 6.17).

6.10 Tool for Gauging Water Flow

The tool is called "raqam" that is in fact a wooden weir to gauge the water level gently passing through it. This tool is made up of a wooden rectangular frame whose one side is marked according to which the level of water can be measured. The level of water in the frame shows the flow rate, and the frame is usually installed in the place where the slope is almost zero and water speed is very low (Fig. 6.18).

6.11 Qanat Working Team

To build a qanat different workers are needed, each of whom is specialized in a particular job. A working team for building or repairing qanats is called "Dast Charkh" which is made up of at least three people including a "Charkh Kesh" who works with windlass, "Lasheh Kesh" who drags the buckets along the tunnel, and "Kolang Dar" who digs the tunnel forward. Three people on a work team are the minimum number. Sometimes two other workers are added to the team; "Gelband" who dumps the spoil in the buckets and "Dalv Gir" who takes the buckets off the hook on the surface and dumps the debris around the well mouth. A working team may include even more people in the long galleries on condition that their number does not disturb their work. "Charkh Kesh" operates the windlass with his hands or feet to send the rope down the well and pull things and people up. "Lasheh Kesh"

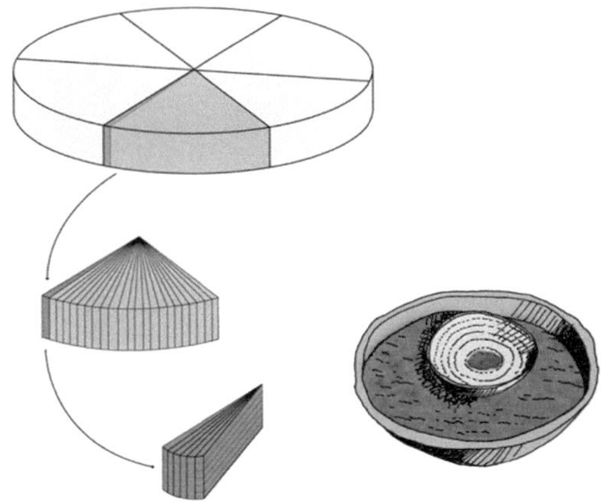

Fig. 6.17 calculating water shares by means of a traditional clepsydra

Fig. 6.18 Wooden weir to measure the water discharge

hooks the bucket of soil to the rope after dragging it all the way from the gallery end to the first well down slope. In the past this person was usually a little boy some 10 or 12 years old, whose small size allowed him to move along the narrow tunnel more easily than an adult. These boys mostly came from poor families whose poverty drove them to send their sons to such a hard menial job. These young people could be promoted to the position of a qanat master if they did not give up and remain in this profession (Fig. 6.19).

One of the qanat masters interviewed in Yazd remembered his childhood in qanat when he used to work as a "Lasheh Kesh". He said he had to work non-stop for 12 h a day when he was 11, because his father had died and his mother and his little siblings made ends meet just with his wage. He remembered those harsh days when he used to drag the heavy buckets of soil while crying out of exhaustion. Nowadays

Fig. 6.19 Qanat working team

hiring under-age boys has fallen out of favor as this practice has been outlawed and more people know about the right of children. Also, in rural regions the economic situation has improved to the extent that they no longer have to engage their children in hard labors. This social transition warrants urgent measures to replace this job with a suitable technology. The qanat masters no longer find small boys to work in the tunnel, so they have to hire adult workers for this purpose and it leads to digging a larger tunnel for them. A larger tunnel means digging and removing more soil at a higher cost, which can put this profession in jeopardy if technology does not find a remedy (Fig. 6.20).

"Kolang Dar" or "Moqanni" is the qanat master who works in the front of the gallery, digging through the soil. Therefore he should be the most skillful worker who knows how to connect a well to another from the bottom by digging a tunnel as well as many other engineering details. A qanat master is paid more than the others and heads the working team. Other workers can become a qanat master one day if they try to learn all the knowledge and skill required.

In the past the workers used to work between 8 and 12 h a day. Two qanat masters took turns working in the gallery once every 4 or 6 h. A qanat master worked in the tunnel from sunrise to noon while the other was busy gathering firewood and preparing meal for the workers. At noon the workers took a break and had their lunch and then began the second shift of their work. This time the second qanat master took the place of the first one in the tunnel until sunset, while the first one was involved in the preparation of food, tools, fire, etc. "Charkh Kesh" and "Dalv Gir" also swapped their jobs during a day, so that one was "Charkh Kesh" and the other was "Dalv Gir" from sunrise to noon and then they changed their places from noon

Fig. 6.20 Promotion of a qanat worker through their hierarchy

to sunset. The hardest job was that of "Lasheh Kesh" and "Gel Band" since they had to do the same work nonstop for 12 h and they did not get out of the qanat even for a break at noon and their lunch was sent down to the tunnel.

Also, the workers' work hour depended on the pace of their work. In case they would contract to dig a particular length of tunnel each day, they could not give up digging until they finished with it. In the past the traditional tools and equipment did not allow the workers to dig at a high pace and they could dig only 2 m a day in a tunnel 60 cm wide and 180 cm high, though they worked round the clock. In fact the pace of excavation correlated with on a variety of factors such as type of soil, type of windlass, number of workers, etc. If they worked in a soft ground, the qanat master could dig 6 m every 6 h, and a double windlass was used to keep pace with the mass of spoil ready to be lifted out. Therefore the workers were able to keep their works in sync with each other, and as a result more length of the tunnel could be dug per day. In case the qanat cut through a hard formation, the workers could rarely dig more than 0.5 m every 6 h.

6.12 Qanat Workers' Overall

The workers used to wear two types of overall, one for working in water production section (WPS) and the other for water transport section (WTS). The overall of the WPS was made up of tanned cow leather smeared with oil, so that the water pouring down on the worker cannot drench the worker. They put on a hat made of the same material. Nowadays the workers are used to wear plastic waterproof overall for the same reason.

In the WTS, the workers used to wear a white textile overall and a hat padded with cotton. The color of their cloth helped them spot one another more easily in dark. Moreover their white cloth could serve as their shroud, since the workers believed that their job is so dangerous that they may be trapped and killed in an abrupt collapse underground while working. Nowadays the workers use the normal overall and safety hamlet with head light.

6.13 Bookan

The qanat workers usually preferred to stay in the place of their work, due to the long distance they had to travel to their homes and lack of transportation. They used to get back home once every two month and they had to stay in an underground hut called "Bookan" all over this time. Bookan was built underground close to the qanat in order to provide the workers with shelter while working in the qanat. A typical Bookan enjoyed a room for sleeping, a chamber for cooking and a place for repairing and sharpening the pick-axes. For example along the qanat of Zarch in Yazd, one can find the vestiges of old Bookans on the surface every 500 m. In addition to

Fig. 6.21 Bookan; underground hut to accommodate the qanat workers during their working season

Bookan, along the qanat of Zarch there are some pits on the ground surface, which are called "Goor" literally meaning grave. The workers used these pits to sleep over night during summer when Bookan was warm and uncomfortable (Fig. 6.21).

6.14 Modern Equipment

Nowadays the modern vehicles and transportation have facilitated the workers' commutation between their works and homes, so they no longer need to stay in their works for a long time. Moreover the new equipment and conditions have eased the labor the workers had to put in qanats. Therefore that many workers and that much labor are no longer necessary, and the traditional job specialization has given in to change. Nevertheless the young men are reluctant to get involved in qanat jobs and rarely replace the old qanat masters who are dying one after another. The shortage of skillful manpower has gripped the qanat projects today, and the quality and quantity of qanat projects are not satisfactory though modern tools such as electric elevator, air compressor drill, electric fan, compass, electric generator, compressor, submerged pump, etc are available which potentially can speed up the work to a large extent.

Reference

Papoli Yazdi Mohamad Hossein, Labbaf Khaneiki Majid, Labbaf Khaneiki Rajabali, Jalali Abbas (2000) Qanat of Ghasabeh Gonabad as a myth. Khorasan Regional Water Authority, Iran

Chapter 7
Qanat Maintenance and Preservation

Abstract When the workers wrap up the construction of qanat, it does not mean that their job is over. They have to keep tabs on the qanat all the time to keep its water flowing out or even increase the qanat water discharge. They have some tools and methods to examine the integrity and health of qanat structure. This chapter takes up the traditional methods to keep a qanat in shape by preserving and maintaining its mechanism such as extending its gallery into water bearing zone, deepening, branching, dredging, lining, etc. In this chapter we also investigate the destructive factors such as earthquake and human induced threats, their impact on qanats and how to keep them at bay. Therefore this chapter focuses on the measures taken to preserve qanat and maintain or increase the qanat water flow.

Keywords Water flow • Groundwater depletion • Tunnel obstruction • Aquifer recharge • Qanat bound

7.1 Measures Taken to Maintain or Increase the Qanat Water Flow

It is likely for water in a typical qanat to dwindle mostly because of a dramatic fall in the amount of annual precipitation, tunnel obstruction caused by collapse or sedimentation or falling objects, water escape from water transport section, frequent droughts, negligence and eventually increasing numbers of tube wells and overexploitation of groundwater. So the qanat masters usually take the following measures to overcome or at least mitigate the problem of water decrease in qanats:

1. Tunnel checking
2. Tunnel cleaning
3. Tunnel extending
4. Tunnel branching
5. Tunnel deepening
6. Tunnel doubling
7. Removing sediments

A.A. Semsar Yazdi, M. Labbaf Khaneiki, *Qanat Knowledge*,
DOI 10.1007/978-94-024-0957-4_7

8. Tunnel insulation
9. Groundwater recharging

7.2 Tunnel Checking

Tunnel checking is called “Ab-Harzi” which means walking along the tunnel from the qanat exit to the mother well in order to control the pace of water flow. This practice may take the workers several days, depending on the qanat length.

They do so in order to control the tunnel and see whether any obstacle is hampering the water flow all over the tunnel. If the obstacle is small, the worker removes it at once by his shovel, otherwise he marks the place so he can get back and remove it with sufficient tools. In the winters when there is more precipitation, the qanats are more subject to crumbling and collapse and accordingly need more “flow easing”. In a nutshell, Ab-Harzi means controlling or checking the qanat gallery. The shaft wells used for this purpose are sturdy and healthy with suitable footholds through which the workers can easily climb up and down the well. Such wells are usually 6 km away from each other. They choose a well every 6 km, and then a worker climbs down the well while the other waits on the ground with a windlass to pull up the soil or his colleague if needed. They walk to the next well 6 km away, a worker underground and the other on the surface, keeping pace (Fig. 7.1).

7.3 Tunnel Cleaning

Some qanats need to be cleaned or dredged every once in a while and some others do not. For example the qanats which run through clay formations should be cleaned very often but those qanats in sand and gravel are not cleaned that often. It is enough

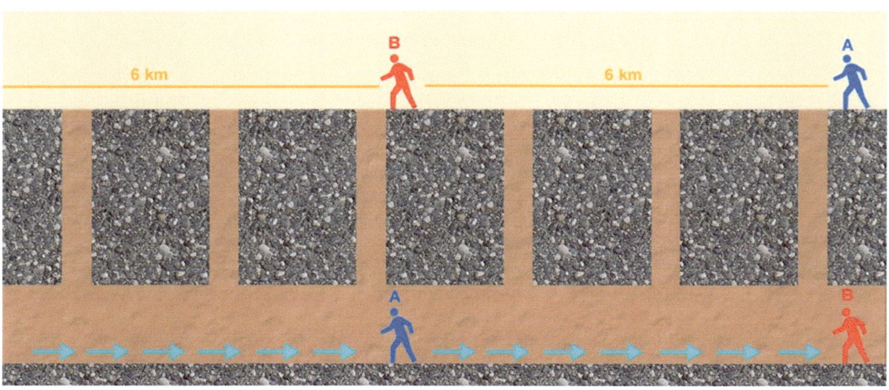

Fig. 7.1 Tunnel checking

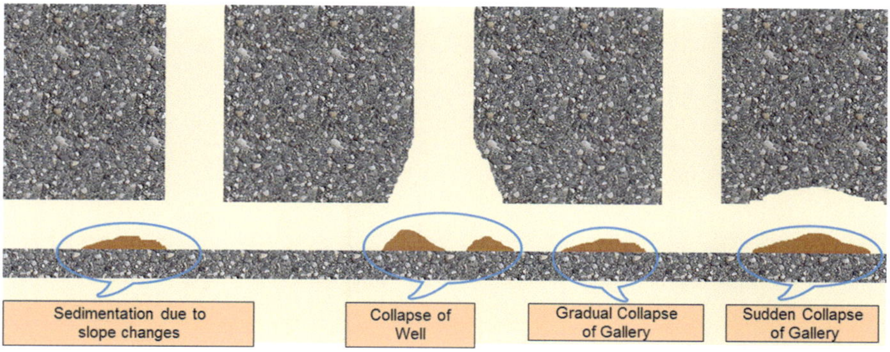

Fig. 7.2 Types of crumbling in some parts of galleries and shaft wells

to ease the flow in the qanats which are not crumbling, but the qanats with any type of crumbling should be cleaned out at least once a year (Fig. 7.2).

Cleaning is referred to as removing whatever object that obstructs the qanat tunnel. The obstruction may be caused by tunnel crumbling or erosion which makes the soil particles build up in the tunnel. Also the tree roots can clog the tunnel if they penetrate the tunnel and get tangled there in large number. In this case the workers have to cut off all the roots and haul them out of qanat. The tree roots can also disintegrate the soil by growing into the porous structure of soil, and can intensify the tunnel crumbling if not removed. Therefore on the earth surface where a qanat is running, the qanat masters do not recommend planting the trees which tend to get their roots to the water sources underground rapidly such as Tamarix, mulberry, walnut, and Haloxylon. Cleaning in water transport section leads to easing water flow and reducing water loss, but in water production section cleaning mostly helps to unclog the tiny holes through which groundwater seeps into the gallery.

7.4 Tunnel Extending

Tunnel extending means advancing into the aquifer by digging the end of the gallery in order to increase infiltration area in the tunnel and get more water. This practice is the most common one to increase water discharge of qanat, though it is only possible and allowed where a qanat does not trespass on the bound of another qanat. The longer water production section, the more water the qanat would drain out, though such a long WPS may lead to a decrease in the discharge of surrounding qanats. In Iran, extending of qanats is normally monitored in order to restrain them from overtaking each other and advancing into the aquifer any farther. This issue is of great importance in the region, because the capacity of the aquifer is limited to be discharged, and the adjacent qanats would run short of water if a qanat nearby would extract water more than allowed. Extending the qanat tunnel through the aquifer can broaden the contact area of the tunnel with water bearing zone on one hand, and

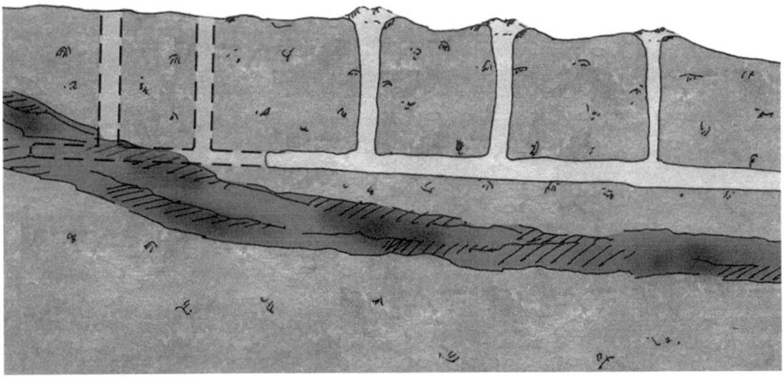

Fig. 7.3 Tunnel extending

increase the height of water column above the tunnel on the other hand. These two factors give rise to more water infiltration into the qanat tunnel. In case "tunnel extending" would be carried out in various directions or in other words in different side branches simultaneously, the discharge of qanat would accordingly multiply for the same reason. Tunnel extending is carried out usually along with digging new wells which are meant to lift out the spoil and ventilate the tunnel. Therefore extending takes place at the end of tunnel which is called "Pishkar" or "Sineye Kar" which usually moves forward into the aquifer. Tunnel extending is called "Pishkar kani" but has some other local names such as "Now Kani" or "Now Kari" (Fig. 7.3).

7.5 Tunnel Branching

Adding more side branches to the qanat main tunnel can increase the infiltration area and accordingly the discharge of qanat. In fact each branch acts independently and eventually their waters join and flow down together. Qanat masters usually prefer the direction of a gully to orient a side tunnel. By digging a tunnel along a gully, they would be able to collect the water already stored in the sediments beneath the gully. Some qanats enjoy more than five side branches each of which has a special name. The side branches are usually named after their founders or the people who managed to dig them for the first time. Apart from the role of side branches in increasing the qanat discharge, they also can expand the vicinity of qanat, which serves as a very important advantage. The side branches spreading around a qanat serve as a buffer zone on which no other qanat can trespass.

7.6 Tunnel Deepening

Tunnel deepening is referred to as "Kaf shekani" which means partially or thoroughly digging out a certain layer of the tunnel floor.

This method can be a substitute for tunnel extending or sometimes these two methods are complementary. Sometimes a drawdown takes place in the aquifer and the qanat tunnel is left out of water. This process has become more common over the past years in the wake of persisting droughts and over-pumping of groundwater. To solve this problem, the qanat gallery should be put in contact with water table once again, otherwise the qanat would remain useless. To do so, it is not always possible to extend the qanat tunnel into the aquifer horizontally, due to some natural and economic obstacles. For example a very hard formation lying ahead of the tunnel may rule out extending, or the tunnel may pass by the aquifer and reach dry zone again if being extended any farther. Moreover each qanat has a buffer zone or vicinity which should be taken into account while being extended. Therefore some qanats are preferred to be deepened – or extended vertically – to avoid all these obstacles. To do so, the qanat gallery is deepened from its exit point to the mother well all the way in order to access deeper layers of saturated area. In other words the floor of qanat tunnel is dug deeper as water table goes down. Deepening of tunnel results in more water seepage into the tunnel and sometimes the amount of water is so much that the workers have to widen the bottom of tunnel in order to better ease the water flow. It should be noted that the tunnel is not deepened more than 1.5 m in most cases, because it does not make economic sense to remove and lift out such a huge amount of soil along a tunnel all the way. It is very important to observe the gradient of the tunnel floor while deepening the tunnel. In case the qanat tunnel is already steep, it is possible to deepen only its water production section while still retaining enough slope for water to flow down. Otherwise the qanat tunnel should be deepened thoroughly to avoid any reverse slope along the tunnel that can hamper the water flow (Fig. 7.4).

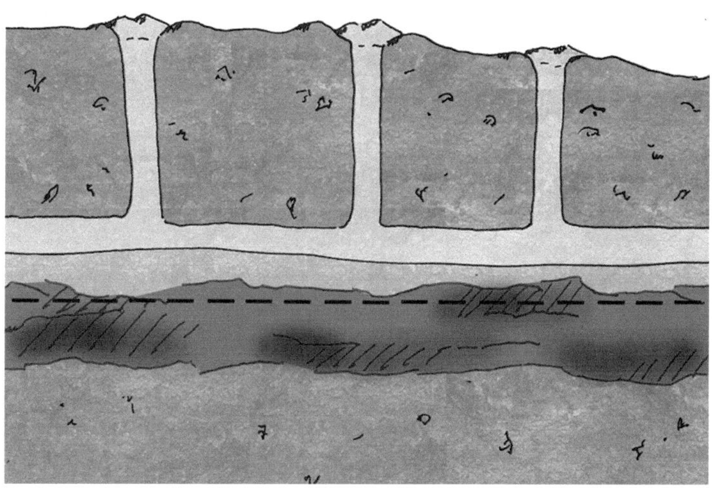

Fig. 7.4 Typical deepening (the *line stipple* is the level to which the tunnel should be deepened to reach groundwater)

Fig. 7.5 Kaval Sargardani

If a part of tunnel is lined with reinforcing hoops (Kaval), the hoops are taken from the old level and re-installed in the new one concurrently with the deepening of tunnel. This practice is called “Kaval Sargardani” which literally means turning the hoops. In this way the workers does not need to lift out the spoil because they can dump it on top of the hoops which have been placed down on the new floor (Fig. 7.5).

We can mention one of our own personal experiences in terms of deepening of qanat tunnel, which was carried out in the qanat of “Sheykh Mamudian”, Erbil, Iraq in 2009. This qanat had long dried up, and we were invited by UNESCO Iraq office to figure out how to rehabilitate the qanat. After studying the qanat, we contended that the best solution was to deepen the old gallery from the third shaft well down. We noticed two pump wells nearby, the first one was at the home of Hatam Kheder (UTM 0447911/4034773) and the second one was at the home of Loghman Kheder (UTM 0447881/4034812). We measured the well depth and the depth to water level and compared them to each other and then to the third well of the qanat. Afterward we came to the conclusion that the water table might be 1.5 m below the bottom of the third well. We hired two workers to dig up the bottom of the well and eventually our prediction came true. To estimate the potential discharge we installed an electric pump in the borehole to pump up a definite amount of water, and then we calculated the time during which the water seepage replaced the same amount of water already pumped. The discharge rate in the borehole turned out to be some 2 l per second which was acceptable, given that more water could infiltrate into a longer tunnel (Fig. 7.6).

After we decided to deepen the gallery to reach water table again, we had to calculate where the new exit point would turn up. To do so, we used an ancient

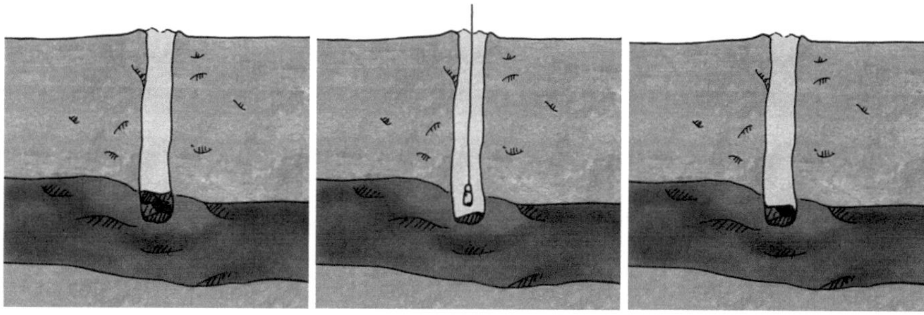

Fig. 7.6 Lifting out water from a borehole to gauge the potential water discharge of a qanat

method (Taraz) to figure out where to dig the new exit point. We held two sticks vertically 10 m apart from each other with a rope in between. We put a level tool on the rope and lowered the rope at one end until the level tool showed that the rope is quite horizontal. Afterward we measured the distance the rope came down on the stick. We made a note and repeated this measuring until we reached the place of the new exit point. It is worth noting that later we called in a topographer to determine the place of the new exit point by means of modern devices to make sure about our measuring. To our surprise he designated the same place as we had found by our traditional tools (Labbaf Khaneiki 2009).

7.7 Tunnel Doubling

If a dramatic drawdown breaks out in the aquifer and groundwater falls considerably below the qanat tunnel, deepening no longer works and the workers resort to the method of doubling. It is worth noting that qanat masters prefer to deepen the qanat tunnel if water depletion is just less than 1.5–2 m below the tunnel floor. Tunnel doubling is called "Tahsoo Roosoo" which means digging another tunnel below the old tunnel and just in parallel with it, so that the new exit point turns up some distance down slope from the old one. In Yazd, Iran the old tunnel is called "Roosoo" and the new one "Tahsoo" and the soil mass between the two tunnels is called "Pol" or bridge in English. The reason why the new tunnel is dug just beneath the old one is to use its shaft wells. The shaft wells are deepened to reach the new tunnel, so the workers do not have to sink new shaft wells from the beginning. Thus the workers economize on time and energy by using the old shaft wells already done (Fig. 7.7).

Sometimes this kind of qanat is mistakenly called two storey qanat which is far away from reality, because in this case the upper tunnel is phased out and the lower one only works. A two storey qanat enjoys two functional tunnels tapping water from two separate aquifers lying at two different levels. For example the qanat of Moon is a real two storey qanat located in Ardestan, Isfahan province, Iran.

Fig. 7.7 Typical tunnel doubling

Water is flowing in its both tunnels without water percolates from the upper tunnel downward. Qanat of Moon runs across the district of Mahal east of Ardestan. Its mother well has long disappeared so that even the locals cannot pinpoint it on the ground, nevertheless this qanat is cleaned every year from its exit point to somewhere near Esmayil Shrine. Near Esmayil Shrine there are two wells close together, which are called Doqoloo – twin – by the locals. One of these wells is 27 m deep which belongs to the upper tunnel and the other well is 30 m deep for the lower tunnel. From this point on the two tunnels enjoy common shaft wells. The two tunnels are 3 m apart with a layer of impermeable clay in between. This qanat is some 2 km long with 30 shaft wells which have been dug every 25–30 m. Each tunnel has a separate mother well, but most of their wells are common. The upper tunnel appears on the surface some distance upslope from the lower one, but their waters join together in a ditch and flow toward the farmlands (Safinejad Javad 2000a, b: 65–71). As shown in the figures below, the both tunnels share a series of access wells around which the upper tunnel turns. The upper tunnel can get access to each well through an opening where the tunnel turns around the well (Fig. 7.8).

Sometimes the workers carry out a combination of deepening and doubling. In some qanats the water production section may be steeper than the water transport section for technical reasons or even mistakes. In this case the WPS is very vulnerable to groundwater depletion, because it does not go deep into the saturated area. After a drawdown in water table, such a WPS is left out of aquifer and the workers have to deepen it in order to hit groundwater again. To do so they deepen the WPS with the same slope as that of WTS until they reach the depth of 1.5 m. After that the workers manage to dig a parallel tunnel beneath the old one rather than deepen the tunnel to such a depth (Fig. 7.9).

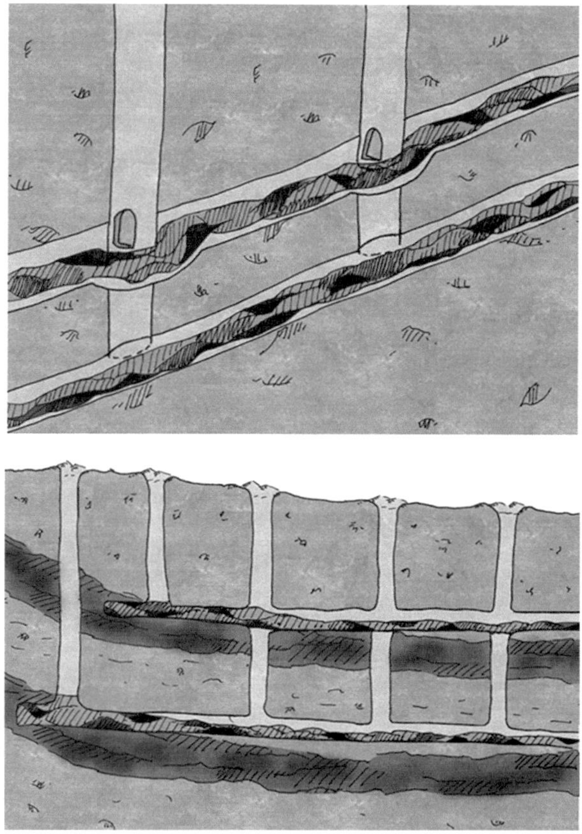

Fig. 7.8 The two storey qanat of Moon with two mother wells enjoying two functional tunnels tapping water from two separate aquifers

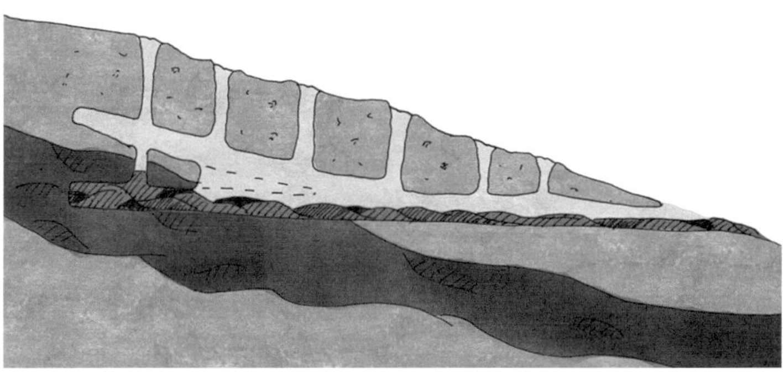

Fig. 7.9 Using tunnel deepening and doubling in combination

7.8 Removing Sediments

In case the water is laden with calcium bicarbonate, physical condition of water and some chemical reactions led this soluble to settle and harden on the qanat floor in the form of calcium carbonate. This sedimentation decreases the infiltration capacity of the gallery and diminishes its cross section which leads to a decrease in water discharge in qanat. Sediments settled on the tunnel floor and sides should be removed and lifted out of the qanat, otherwise the tunnel may be clogged over time. These sediments are called "Shaah" which should be taken off the tunnel wall very carefully such that a thin layer of the sediment would be left on the wall, because it serves as a natural lining which curbs water percolation. Therefore the workers prefer not to manipulate the sediment as long as its build-up does not interfere with the qanat discharge. When the sediment would accumulate so much that it obstructs the water, the workers manage to scrape a layer of sediment off the tunnel wall such that the water current would be eased and on the other hand a layer would be remained to insulate the tunnel.

7.9 Tunnel Insulation

The water transport section is insulated in order to stop water percolation. The discharge of a qanat depends on the amount of water wasted along the tunnel in water transport section. The longer the water production section, the more water would percolate. Also type of soil correlates with the amount of water loss, since the water percolation in sand and gravel is more than that in clay. In the shallow parts of water transport section, the workers sometimes dig the tunnel open like a trench and shore it up with brick and cement all the way, and then refill the trench up to the surface. To insulate the tunnel floor in the water transport section, the workers also used to tread clay mud into the cracks and gaps on the floor. This practice is called "Lamal Kardan"; the worker spreads some clay mud on the tunnel floor and walks on it so the clay goes into the cracks through which water may escape.

Also the workers fill a sack with a mixture of straw and soil and drag it behind themselves along the tunnel, while someone else pours some soft clay down a well nearby. The sack helps the clay better mix with water and penetrates the holes and cracks on the tunnel floor. Bentonite is also used to insulate the tunnel floor. To do so, they mix Bentonite with qanat water, 25 kg of Bentonite in every 100 m of tunnel, so that Bentonite settles on the floor and stops water percolation. Nowadays the water transport section is insulated sometimes by spreading the sheets of plastic on the tunnel floor all the way from the exit point to the border between WTS and WPS. These sheets are 120 cm wide.

In the past, the workers used to coat the tunnel floor with a thin layer of clay mud and then put the sheets of canvas on it. The edges of canvas were fixed on the tunnel wall with pieces of mud, and then they coat the canvas with another layer of clay

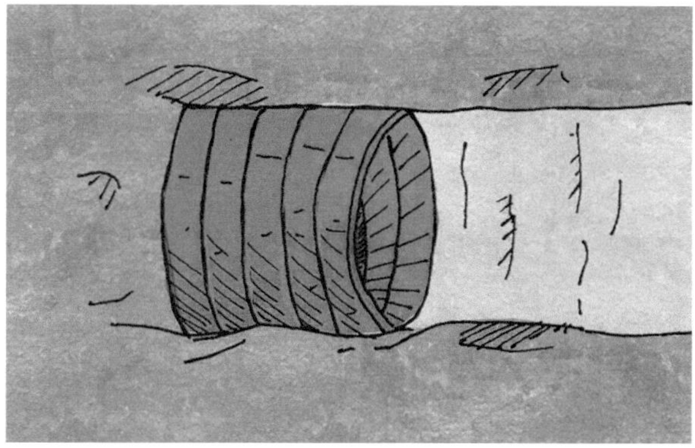

Fig. 7.10 Hoops installed in a qanat tunnel side by side

mud which could fill up its porous texture and get it waterproof. In case they use the above mentioned methods to insulate the tunnel, they should look after the qanat more carefully so the qanat would not need cleaning very often otherwise the insulation would be damaged soon (Fig. 7.10).

Eventually the other way to solve the problem of water percolation is ceramic or cement hoops (Kaval) which are installed along the gallery one after another the same they are used to reinforce the tunnel against collapsing. In the water transport section, the space between the hoops is filled up with cement to make them seamless so the water finds no way to escape. It is worth noting that professional qanat masters are usually reluctant to use such hoops for the purpose of insulation or reinforcing, unless they have to do so, and they cannot find any other solution. Sometimes they call these hoops "Gom" which literally means "Lost", because they believe that a tunnel lined with hoops is going to be lost. A qanat tunnel should be dynamic and able to move horizontally and vertically to keep pace with water table. A tunnel lined with hoops is very difficult to be dug and deepened if groundwater drops below the tunnel some day (Fig. 7.11).

7.10 Groundwater Recharging

Groundwater recharging is referred to as all the measures taken by people to replenish the aquifer. As far as we know, these measures fall into four types: artificial recharge dam and pond which is respectively called "Band" and "Goorab", underground dam which is built to store groundwater in qanat, qanat valve which is used for the same purpose and eventually injecting flood water into abandoned qanats. We describe these four types as follows.

Artificial Recharge Dam it is a traditional dam built for harvesting seasonal runoffs. During the rainy seasons, the runoffs flowing down the surrounding hills were directed to the dams. In the dams water built up and gradually percolated the earth into the aquifer which fed the surrounding qanats. There still exist many traditional

Fig. 7.11 A qanat worker dragging a sack of soil and straw to level the mud and insulate the tunnel floor

recharge dams in the region of Ferdows, South Khorasan province, Iran. This region is completely out of permanent surface streams except for some seasonal runoffs which originate from the eastern and north eastern mountains and flow toward the less elevated lands between Boshrooyeh, Nignan and Bajistan. Insufficient precipitation made the farmers resort to building some earthen dams along the gullies to trap the seasonal runoffs in order to replenish the aquifer so their qanats can get through the dry seasons. Chenjeh is one of those gullies which is 90 km long and passes by the town of Ferdows. Also the gullies of Boroon and Sorkh-Ow are among the most important gullies in the region. Most of the traditional recharge dams are constructed upstream of the plain in order to make the most of the seasonal runoffs (Fig. 7.12).

The dams are preferred to be built on the grounds with 2–4 % gradient which is optimum to reduce the soil erosion. A typical dam consists of the following components:

1. Divareh or the wall of dam: in Khorasan province, Iran the wall of dam is usually built with piles of soil and sometimes in combination with stones. The wall is usually 1–2 m high, and in some areas a mortar of clay and lime is also applied to shore up the dam.
2. Sartag: a ditch that brings water to behind the dam is called "Sartag" which derived diagonally from the main stream in the gully. The structure and size of Sartag depend on the surface slope and the size of dam.
3. Dam exit: dam exit is a little bit lower than the top of dam wall, through which water can overflow. In case there would be much water behind the dam or the wind would make waves battering the dam, the excess water spills out of this exit before the dam gives way. The exit is built just across from Sartag, on the opposite side of the dam.

Fig. 7.12 Artificial recharge dam

4. Dam hut: the smaller dams needs more care, so a hut is built close to them. The dam is recharged when a torrential downpour takes place, and those who are to keep a watch on the dam need a shelter against rain and cold. This hut is circular some 3 m in diameter, with an adobe dome roof which is 2.5 m high.

Artificial Recharge Pond in the flat area, to trap a runoff and make it seep into ground, a relatively big pond was dug out and the spoil was accumulated around it mostly around its downslope. This structure is called "Goorab" which literally means the grave of water (Fig. 7.13).

In the region of Taft, Yazd province Goorab or groundwater artificial recharge ponds has once played a crucial role in water supply and accordingly in agricultural production systems. Such dams were built upslope from the qanat mother well, where called "Boxum" in local dialogue. These ponds were built upslope from the qanat, which was usually located in the territory of another village. This issue became a matter of cooperation between the two villages, as they both could benefit from this relationship. Sometimes the upslope village took advantage of these ponds by growing some crops in the ponds after their water went down and left suitable moisture in soil. On the other hand the downslope village could benefit from these ponds by replenishing the aquifer and increasing their qanat discharge. Unfortunately at present the vestiges of those traditional ponds have disappeared from the landscape, and were replaced by different constructions or cultivated lands now irrigated

Fig. 7.13 Goorab and its role in replenishing groundwater

by means of tube wells. For example the qanat of Chahoke Nir in Taft once enjoyed four recharge dams which came into the possession of the residents of the village of Pandar after the Islamic revolution of Iran. The residents of Pandar changed the use of the dams though they had an important role in replenishing the aquifer on which their qanats used to rely. The ponds were located upstream of the qanat mother well in the lands of Pandar north of the village of Chahok. In the past the residents of Pandar owned some shares of the qanat of Chahok, so they not only tolerated the presence of those ponds in their territory but also helped to fill them with the seasonal runoffs. Later the residents of Pandar sold out their shares of qanat and were no longer beneficiary of the ponds, so they were encouraged to demolish the ponds and annex them to their own farmlands. The annihilation of the ponds led to a dramatic decline in the discharge of the qanat of Chahok (Fig. 7.14).

After many traditional recharge dams vanished, in 1989 the Iranian government came to realize the importance of artificial groundwater recharge and took some measures in order to resurrect the dams but in a modern way. For example Yazd Regional Water Authority conducted some study projects on the feasibility of implementation of artificial recharge dams across the province. These studies resulted in construction of 18 earthen recharge dams with a total budget of 2,525,252 USD by the year 2001 in order to harvest the seasonal runoffs and replenish the groundwater reserves.

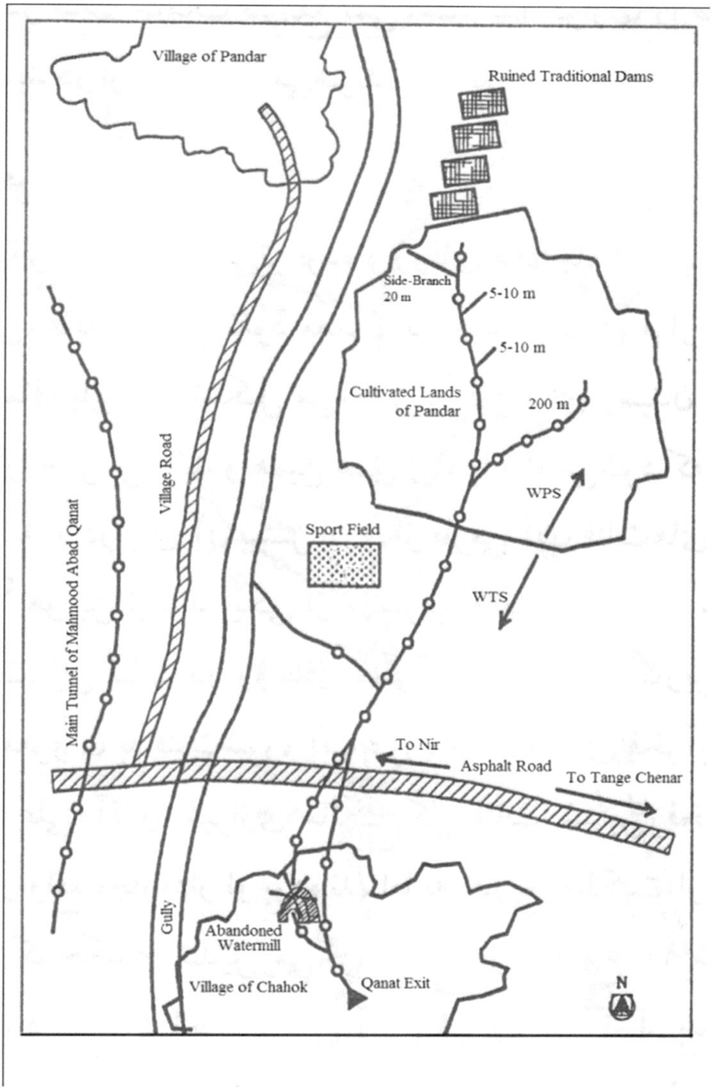

Fig. 7.14 Map of qanat of Chahok and the position of its traditional recharge ponds (Goorab)

Is not it true that maintenance of the traditional dams would cost much less money? This is another example of the follies of modernity. Any way through the implementation of these new dams, 17 million cubic meters of seasonal runoffs were harvested across the Yazd province and gradually injected into the aquifer. One of these dams has been built of stone. This dam which is located upslope from the village of Nir, serves to trap the runoffs flowing down the valley and makes them seep into the aquifer. This dam is 140 m long and 5 m high with a water capacity of

100,000 cubic meters. Nowadays research and study on artificial groundwater recharge are catching on in Iran, which can pave the way for preservation of the aquifers across the country. One of these studies is entitled "evaluation of artificial recharge and its quantitative impact on groundwater level" which suggests a computer program for tracing the effect of artificial recharge on groundwater. This study also sheds a light on the effects of such factors as surface slope, geological and soil conditions, type of aquifer, climate, etc on the efficiency of the artificial recharge. Also the university of Ahvaz Shahid Chamran conducted a project entitled "application of computer modeling in the management of artificial groundwater recharge in the plain of Ghorveh". This study gives insight into the correlation between over-exploitation of groundwater reserves and environmental economic crisis in the region, and the capacity of artificial recharge projects to mitigate the problem (Salmasi Farzin 1997). Such studies and projects are making up for the lack of traditional recharge dams, but it should be noted that the function of traditional dams was not only to replenish the aquifers. The traditional dams demanded an extensive social cooperation, paving the way for a social convergence. Moreover the farmers used to cultivate some crops in the land behind the dam when it was out water. The traditional recharge dams were an important component of the human ecology in the desert environment, which unfortunately died out along with its heritage of social values.

Aquifer Recharge Through Abandoned Qanat Tunnel A qanat tunnel may be tens of kilometers long with several side branches. In the wake of a boom in the number of tube wells, in some regions the qanat system gave in and fell into decay, but their empty tunnels can still be used for recharging groundwater. There are some experiences in pouring surface runoff into the qanat shaft wells and injecting it into aquifer this way (Fig. 7.15).

If there is an inactive qanat in the neighborhood of an active qanat, the inactive qanat can be used to inject surface runoff into the aquifer which feeds the active qanat. For example in Meybod, Yazd province a seasonal runoff was diverted into

Fig. 7.15 Using an abandoned qanat tunnel to recharge aquifer

the inactive qanat of Hasan Abad, which later led to an increase in the discharge of the qanat of Kharzar at a distance of 6 km.

Underground Dams In some qanats, a very simple method is used to cease the qanat flow in winters by blocking the tunnel when the farmers no longer need water for irrigation. This way the water is stored in the porous sediment, increasing the qanat flow when needed. For example in some villages of Kashan, Isfahan province, they put some big wads of rag in different points of the tunnel in qanat water production section in order to stem the flow. This practice is possible in the tunnel wherever the soil structure is sturdy and is not vulnerable to water build-up, otherwise it collapses. Thus the qanat gets blocked every year in autumn until late winter (Safinejad Javad 2000a, b: 228). Also there is a qanat in Shahr Babak, Kerman province which enjoys an underground dam in order to store water. Some 100 m down slope from the border between water production section and water transport section, a wall has been built across the tunnel with brick and lime, blocking the tunnel from bottom to top all the way. In this wall, there are several openings which were kept close during winter when people do not need water, so water accumulated behind the wall. The openings are made in a row just in the middle of the wall from top to bottom. Each opening is in fact the space of a missing brick, so when they want to close the opening they put the brick back into it and fix the brick in place with clay mud. This technique is feasible only in the qanats which enjoy a firm and sturdy structure, otherwise the tunnel would be soaked in water and its roof and sides would start collapsing. The workers never build the dam far away from the WPS just in the WTS, because water percolates the earth and escapes from the qanat access. Another example of this dam can be found in the qanat of Vaghf Abad in Yazd, though the most important groundwater dam in Iran is considered the dam of Vazvan qanat in Isfahan province, about which a book has been authored by Javad Safinejad and Bijan Dadras (2000). The total length of this qanat is 1.8 km from mother well to exit point. 1.2 km down slope from its mother well an underground dam has been built across the tunnel with brick, stone and Sarooj[1] at a depth of 16 m from the surface. The dam is not uniform from bottom to top, but its bottom is thicker than its top and the dam is crescent in shape. There are five outlets made into the dam wall, with different distances from each other all in a row from top to bottom. Due to the water pressure stored behind the dam, the outlets were built in the shape of a funnel with a narrower entrance behind the dam and a wider exit on the other side. The two sides of each outlet are different in size by 40 %.

The dam outlets were shut in early autumn for 4.5–5 months to store up to 270,000 cubic meters of water in the qanat. When the time came to release water, the outlets were opened respectively from top to bottom. They opened the upper outlet and waited for 25 days until the water level behind the dam went down, and then they opened the next outlet until they reached the last outlet which was called

[1] Sarooj is a traditional mortar formerly used in Iranian architecture. To make Sarooj it was common to mix clay and lime together by a ratio of six to four. The workers kneaded this mixture in turns for 2 years and then added some ash and herbal fibers to it to form a very sticky and firm mortar.

Fig. 7.16 Typical underground dam (*Left*: profile, *right*: front)

"Daricheye Takht" which means the outlet of floor. Recently some modern projects have been carried out in Iran in order to store qanat water during winter, though their technical principles are the same as that of traditional methods. As an instance a project in Isfahan province is pointed out as follows (Fig. 7.16).

Qanat Valve This method serves the same function as described above, except for the technique used here which is modern. One of the technical disadvantages of qanat system is the constant flow of water even during winter when the farmers do not need it. To solve this problem some projects have been carried out in different parts of Iran in the qanats whose structural conditions allowed. For example the qanat of Rabi Looshab in Meyme, Isfahan province was selected for installing control valve in its gallery. This qanat is 1600 m long with a 16 m deep mother well. Its shaft wells have been sunk some 50 m away from each other, and the water control project was carried out at a point some 100 m down slope from its mother well. To do so, the tunnel has been blocked by a concrete wall in which two 16 inches valves have been embedded. One of the valves is operational and the other is spare. When water is needed the valve is open otherwise it is close. Form this point upstream, the qanat gallery cuts through layers of shale which is resistant to water build-up, but from this point downstream the geological formations are crumbling and subject to collapse. Every year on 11 November the valve is turned off and the water reaches a depth of 10 m in the mother well in 15 days. After 15 days, water level no longer goes up in the mother well and it remains steady, but water penetrates all the gaps and cracks in the layers of shale, stored there for 4 months. The valve is turned on around 4 April to release the water. It should be noted that some water escapes into the tunnel and flows down a few meters down slope from the valve at a rate of 8 l per second, but the rest of the qanat discharge – 12 l per second – is still stored as mentioned.

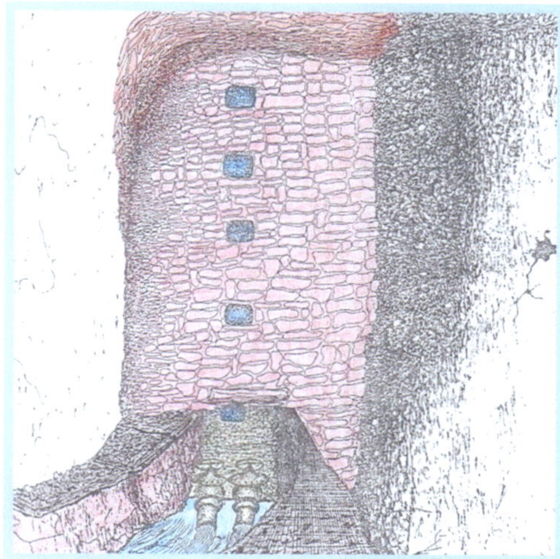

Fig. 7.17 Typical qanat valve (Source: Javad Safinejad 2000b)

On 4 April when the control valve is turned on, water pours down at a rate of 30 l per second, and after 15 days the water level in the mother well goes down to the tunnel floor. From then on water starts seeping out of the layers of shale into the tunnel at the same rate during spring, but the discharge gradually drops to 20 l per second. Therefore, storing water in this qanat results in additional 10 l per second of water during spring and 5 l per second during summer, which could not be achieved without the control valve. This additional discharge has led to cultivation of 15 ha of land which has been added to the previous area of cultivated lands. Moreover this project could put an end to the problem of water shortage in the village whenever a drought breaks out (Azad Mohammad 2000: 113–117) (Fig. 7.17).

7.11 Measures Taken to Preserve Qanat

Traditional qanat masters have invented some techniques and methods to protect qanat against such natural hazards as earthquake and human induced threats like trespassing on qanat boundary. These measures are taken to retain the qanat water flow in the face of those hazards or threats.

7.12 Impact of Earthquake on Qanat

The structures built by human are not spared from earth quake, and qanat is not exceptional. Especially where qanat gallery cuts through a crumbling formation prone to crumbling, the qanat would be more likely to cave in if an earthquake breaks out. There are also some historical records which all bear witness to the vulnerability of qanat to earthquake. If an earthquake strikes the qanat and its tunnel collapses afterward, the tunnel would be obstructed and water would build up there and recede, so that a considerable part of qanat tunnel would be submerged and soaked in water and then tumble in ruin. Nevertheless an earthquake can also have a positive side for qanat as it may lead to an increase in the qanat water discharge. Earthquake can change the structure of soil and new cracks may spring up and serve as underground conduits through which aquifer can be drained out into the qanat tunnel. There are some examples of such qanats whose discharge increased in the wake of a devastating earthquake. Here we use the qanats of Bam, Iran as example to better clarify how earthquake can affect qanats. In December 2003 an earthquake with the moment magnitude (Mw) of 6.6 struck Bam and the surrounding Kerman province of southeastern Iran. This quake led to the destruction of some qanats in the region and on the other hand diverted groundwater to some other qanats, resulting in an increase in their discharge. Some qanats collapsed in the wake of the earthquake and their water was obstructed in their gallery. As a result the water drew back and was diverted to such qanats as Ghasem Abad that showed more discharge after the quake. Some qanats were repaired immediately, and re-construction of some other qanats was put off until the ruins of homes and streets were swept up.

The earthquake kicked off a change in the qanats of Bam from two different standpoints as follows:

1. Structural change: in the traditional manner, the qanat workers used to remove the debris from the tunnel by working underground, but after the earthquake this job was done by using excavator machines which opened the tunnel. The relatively shallow qanats of Bam made it possible for the excavators to dig the surface down to the qanat tunnel like an open trench. The excavators were not limited to the direction of the qanat tunnel running underground. In other words the excavators did not always dig on the top the qanat galley, but they preferred to dig wherever the earth reliefs favored digging. Therefore in some qanats, some parts of the gallery were abandoned and its direction changed. For example in the qanat of Abbas Abad, the upper part of its gallery was forsaken during its re-construction and the qanat changed in length. Any change in the direction of qanat may bring about some social disorders. In traditional water management systems, the closer the farm or orchard would be to the qanat exit, the more expensive and precious the farm. Now any change in the location of the exit point can stir up social disagreement. The trenches dug by the excavators had to be lined and covered with concrete and then refilled with soil. As mentioned, a tunnel with such a sturdy hard floor loses its flexibility to the groundwater fluctuations. In some cases the tunnel should be possible to be deepened to hit the depleted water table again, and it is almost impossible to do so in the concrete tunnels. Moreover a technical mistake aggravated the situation. For example the qanat of Qanbar Abad – Sheykhi collapsed and obstructed at a point located in the garden of Abbas Khan Ameri. From this point down the tunnel was dug by an excavator to remove the obstruction, forming a 120 m open trench with concrete lining whose floor is about 1 m higher than that of tunnel itself. As a

result water comes out from the tunnel but cannot easily enter the open trench because of the aforesaid height difference. Water build-up in the tunnel resulted in more cave-ins and new destructions in the tunnel.

The critical situation did not allow making use of the experience and expertise of the traditional qanat masters. Nevertheless in some areas the topographical condition is not suitable for using machinery to repair the qanat tunnel and release the water, and the only solution is to practice the same traditional methods. As an instance qanat of Kohneh ended up in such situation. This qanat was ranked among the most important qanats of Bam, whose water was rationed during an 8 day rotation of irrigation. In 2005 each hour of water share in this qanat was bought and sold at a price of 2288 US dollars. The earthquake took a heavy toll on this qanat by destroying several parts of its gallery. After the quake by common consent, a project for rehabilitation of this qanat was launched from the exit point up to the mother well. A distance of 100 m from the exit point toward the first well was excavated by machinery as an open trench with a depth of 10 m. From that point on, some qanat masters from the city of Yazd managed to repair the gallery in a traditional manner. They went ahead until near Bam municipality where two sever collapses obstructed the gallery. The workers climbed down a well just between the two collapses and managed to remove the both in two directions at the same time. The workers did not find any more obstructions in the gallery until a place named Helal Ahmar from where to Golestan Cold Storage there were only some slight collapses. After removing those collapses water resumed flowing down the gallery, but the gallery needed more work beneath the last wells near its mother well to be completely rehabilitated. Bam organization of agriculture contracted a traditional qanat master named Avaz Karimi to clean this part of the gallery and finish the project.

2. Managerial change: before the earthquake every qanat in Bam has a council which traditionally managed water division and ownership. For example Mr. Gholamreza Biglari was once the council head of the Ghanbar Abad Sheykhi qanat, who handled the ownership documents of this qanat. In the wake of the earthquake all the traditional ownership documents went missing and the qanat management system fell into the hands of such governmental organization as Jahad Keshvarzi who did not know too much about the intricate mechanism of traditional water management systems in a multidisciplinary context. Such interventions caused a systematic managerial disorder which demands a multifaceted investigation to be fixed.

The cost of labor was on the rise, which had something to do with an increasing demand for manpower in order to rehabilitate the ruined qanats. In 2005 Avaz Karimi was paid 22 USD per each meter of removing debris from the gallery. This amount included 13 USD per day for the wage of compressor drill operator, 8 USD per day for the worker who loaded the spoil into the buckets, 5 USD per day for that who dragged the buckets and hooked them to the rope, 22 USD per day for the tractor driver who lifted out the buckets, and eventually 8 USD per day for that who unhooked the buckets from the rope and dumped the spoil around the well mouth.

Therefore the more spoil Avaz Karimi's workers could lift out from the qanat, the more money he made. In the same year, extending a qanat tunnel into aquifer was more costly than cleaning. The cost of extending varied from qanat to qanat, given the type of soil through which the qanat cuts. This cost varied from 85 to 572 USD for each meter digging.

One of the consequences of the earthquake was the very presence of governmental or non-governmental aid forces who came to the region to help rehabilitate the damaged qanats. Their presence had a deep impact on the qanat techniques and even terminology in the region. For example before the earthquake "meter" as a unit of length was never used by the qanat workers in Bam, but they had their own unit called "Baghal" which equaled 175 cm. After the earthquake the unit of Baghal sank into oblivion and was totally replaced by meter in the qanat workers' terminology.

7.13 Safeguarding the Bound of Qanat

All the traditional qanat masters say about the bound of qanat can be summarized as follows. The bound of qanat falls into three main types which are physical, hydrogeological and qualitative bounds. These bounds are of great importance when it comes to the preservation of qanat and its managerial plan. In a nutshell, physical bound is referred to as a certain area on the surface around the access wells and the qanat tunnel, within which all the activities harmful to the structure of qanat are not allowed. These activities vary from cultivation to construction. Hydrogeological bound pertains to the preservation of the aquifer that feeds the qanat, and the determined area restricts any kind of groundwater extraction which would lead to the depletion of the qanat discharge. Qualitative bound ensures the quality of qanat water and is related to an area around the qanat within which no sewage or any contaminating source is permitted. This bound is determined based on such factors as soil fabric, type of geological formation, porousness of soil, liquid speed though soil, hydraulic gradient, qanat depth, etc.

In the qanat workers' terminology, bound of qanat and bound of well are two different concepts. Bound of well is considered a circle around the well mouth between 2 and 6 m in diameter. To specify the bound of well, a worker sits down in the well while digging it, after the well reaches the depth of some one meter. Afterward the worker musters all his force and throws his pix-axe back over his shoulders. Wherever the pick-axe lands it is considered the far end of the well vicinity which should be observed by all people. That is why the bound of well is called "Kolang Andaz" which literally means "throwing pick-axe" (Fig. 7.18).

In the bound of well no one is permitted to cultivate crops, build homes, irrigate fields or any activities which can potentially damage the structure of the qanat and cause the wells to collapse. Moreover if a well deteriorates and lies in ruins, its surrounding should be unoccupied so it would be possible to dig a new well nearby. Nevertheless over the past decades inattention to the bound of qanat wells has been

Fig. 7.18 Traditional method to determine the shaft well bound

on the rise, which is in fact associated with the developmental projects neglecting such traditional issues as qanats. For example in the village of Nasr Abad, Khorasan province a housing company proceeded to build a compound with no attention to the qanat wells which later turned up just beneath the homes. Some of the wells are just under the living rooms, impossible to be re-opened for the purpose of cleaning and rehabilitation. On the other hand this qanat should be cleaned occasionally to keep alive, otherwise the qanat is doomed to annihilation.

Contrary to the bound of wells, the bound of qanat tunnel mostly pertains to the hydraulic vicinity of qanat in which any water extraction may have negative impact on the qanat discharge. The bound of qanat depends on a variety of factors including the surrounding geological formations, type of aquifer, proximity of other groundwater sources, depth of gallery, etc. In a nutshell, groundwater flows slowly through the porous sediment and it tends to move toward the area where there is a difference in the hydraulic potential. This situation occurs where a source like a qanat or a tube well is discharging groundwater. Therefore if a well or another qanat exists upstream from a particular qanat, groundwater would be intercepted. Traditional qanat masters say that such a well or qanat cuts the vain of the lower qanat, it means that the upper sources harvest the groundwater flow so the lower qanat cannot get enough water. To prevent any interference between the adjacent qanats, there are some local regulations which ensure their sustainability. In Taft region, Yazd province, the plain qanats which run parallel at the same depth can be extended into the aquifer upslope only in two diverging directions. In other words, the workers are not permitted to do any activities such as extending or branching in the distance between the two qanats. In case one of those qanats is deeper than the other, extending is permitted only on condition that the shallower qanat would be extended too, twice what the deeper one would advance. For example if the deeper qanat would be extended 100 m into the aquifer, the shallower one would advance 200 m. In the mountainous areas, there are stricter rules in terms of qanat bound, because the aquifer is more limited and shallower and the qanats run closer to each other. In such areas the workers can rarely dig side branches since they may easily trespass on the vicinity of the other qanats. In the vicinity of qanat it is totally prohibited to dig a well, unless the qanat gallery enjoys a relatively steep slope (about 2 or 3 %) and the well keeps enough distance from the qanat. The qanat masters believe that such a well can be allowed at least 1500–2000 m away from the qanat tunnel where the water transport and production sections border. This is a safe distance which does not affect the qanat discharge. To settle any disagreement between the owners of qanat and well over the location of the aforesaid border, they use a special technique named "Zenama" which literally means "Seepage Indicator". Sometimes the owners of qanat take a particular spot as the border between water transport and production sections, whereas the owners of well like to designate a place more upslope as the border so they would have a better chance to exploit more water. Zenama serves to settle the dispute of this kind. A worker who is mutually agreed upon goes down the qanat and makes a cut into the tunnel wall just where believed to be the border by the qanat owners. This cut – Zenama – is some 60–70 cm long and 15 cm deep. The cut is made only 10 cm above the water level flowing down the tunnel. The worker leaves the cut and waits for 24 h, and the day after he returns to the cut to see whether or not any water has accumulated in it. The presence of water in the cut proves that here is not located in the dry zone or water transport section but it is in either water production section or at least the border between.

In some cases, the earth features such as hills can mislead the qanat workers and cause them to trespass on the vicinity of other qanats inadvertently. According to the qanat masters, the qanats running down the two opposite sides of a hill can freely be extended in every direction because their feeding aquifers are independent and separate. In reality this idea applies only to the elevations which resulted from geological stratification bringing about two opposite hydraulic gradients, but the story is quite different in terms of other hills like sand hills. Such elevations have nothing to do with geological stratification and the aquifer beneath them is integrated and seamless. The qanats draining such an aquifer may trespass on each other if they are extended recklessly, though they lie seemingly on the two opposite sides of a hill.

Apart from the conservation bound of qanat described above, there are some other concepts regarding qanat bound. For example Qanat bound from canonical point of view is still in favor. Religious scholars have determined a specific length as qanat bound. According to historical resources, a strong earthquake struck the city of Far'aneh in Khorasan about 1000 years ago such that a number of qanats in the region were devastated. After rehabilitation, dredging and reconstruction of affected qanats due to high number of qanats, their little middle distance and assaults to the bounds of qanats by other qanat owners serious conflicts arose among people. People referred to religious scholars to solve the problem. Marja[2] of the time gathered and appointed a number of religious scholars from Iran and Iraq and required them to comment on the issue of qanat bounds and its legal aspects. The reviews of

[2] Literally means "Source to Imitate/Follow" or "Religious Reference". It is the label provided to Shia authority, a Grand Ayatollah with the authority to make legal decisions within the confines of Islamiclawforfollowers and less-credentialed clerics. http://en.wikipedia.org/wiki/Marja'_(Islamic_law)

this group is collected in a book called "Al-qani" which surveys many of the issues related to the bound of qanats and has been referred to for many centuries.

Haseb Karaji about canonical bounds of qanats believes that: "Normally the religious sentences related to qanat bounds are referred to since Sharia is more trusted than else. When a qanat is excavated in a permissible land and one aims to excavate another qanat for him/her in its neighborhood, the owner of first qanat cannot hamper the person from doing so provided that the distance between the mother wells of the qanats is not less than 1000 Zera[3] and the qanats are of the same depth." In this way Karaji quots the length of 500 Zera as the canonical bounds of qanats from each side and if each Zera is considered about 1 m then canonical bound of a qanat would be about 500 m from each side. Karaji also has considered regional soil condition, type of aquifer, the way of aquifer recharging, situation the qanat locates (mountain, domain or desert) etc in his discussion on qanats bounds and has emphasized the significance of judges' expertise to determine qanat bounds. He believes that the experts in this field have to know soil and be experienced not to fumble or make mistake in their judgment. As an example, he believes that qanat bound in a hard soil is less than that of one in a loose soil and the harder the soil of a qanat, the shorter the length of the qanat bound down to 40 Zera as a minimum. Determination of this bound is conditioned to the expert being an agrologist. He mentions that whatever he has stated about determination of qanat bound is said based on estimations and thoughts.

According to our interview with Mr. Mohammad Reza Fayyaz one of the qanat masters in Yazd, the canonical bound of qanats is considered 2500 Zera from each side. He emphasizes that the qanats in one another's neighborhood within this limit must have the same depths. If the depths of the qanats are not the same, 1000 Zera is added to the bounds for any meter difference in the depth of qanats in order to prevent any bilateral harm to any of the qanats. Mr. Ali Moqani Bashian a qanat expert states that canonical bound of a qanat is 1000 m in loose soils and 500 m in hard soils. Determination of qanats' bounds has been one of the authorities of the governors and no one was allowed to trespass on those limits. These days, however, the canonical bounds are not observed.

Qanat bound from quality point of view is also of great importance. The purpose of defining a qualitative bound is to prevent any sort of microbial or chemical pollution from penetrating the qanats. Extreme expansion of mega-cities and development of human communities have led to a drastic increase of urban sewage and every moment huge amounts of swage is produced and flows. Lack of a sewage collection system or a proper place to dispose sewage safely may result in contamination of groundwater resources. Using variety of chemicals as fertilizers or pesticides by farmers can be another source of groundwater pollution too. Industrial sewage if not managed in a proper way is another source of underground water contamination. Other than above-mentioned items, expansion of mines and their

[3] Zera' is the plural form of Zar' which is a length measurement equal to 1.04 m. http://www.loghatnaameh.com/dehkhodaworddetail-589562f8f0af48088da685ad50f7f9e9-fa.html

washing operations, hospital wastes and respective latex and urban wastes can be other sources of groundwater contamination. Therefore, prevention of penetrating the mentioned contaminations to groundwater resources and qanats are considered under the title of observing qualitative bound of qanats. As is observed, the qualitative bound is not a function of distance and it is very probable that contamination in long distances away can penetrate qanats gradually and contaminate them. Since the establishment of urban and rural sewage systems, sewage treatment systems, standard disposal of wastes, treatment of industrial swage and finally observing the standards respecting fertilizers and pesticides not only are effective to develop health and sanitation conditions of society but also will contribute to preserving clean water and qualitative conservation of qanats. Therefore, in order to preserve the qualitative bound of qanats all environmental standards respecting pollutions, sewage and wastes have to be observed.

Qanat bound from legal point of view is the last concept that should be noted. Harim is an Arabic word and its lexical meaning is "Proscription". It implies that some of actions and activities within this bound are prohibited or limited. In order to clarify such a bound, all of the laws touching upon qanat in a way have been surveyed which hereby will be discussed. We begin the discussion with civil laws and some of its articles dealing with properties' bounds:

Article 136 – Bound is part of surrounding areas of a territory, qanat, slough and suchlike which is essential to maximize utilization from it.

The implication of this law for the qanat bound is that: We should specify a bound for utilization, repair and maintenance of qanats and their utilization cannot be conducted suitably, provided that this bound is not indicated.

Article 137 – Well bound for drinking and cultivation purposes is 20 and 30 Gaz respectively.

This article refers to the wells dug in the past for drinking or agricultural purposes and their water was extracted by windlass and bucket. These wells enjoy 20 and 30 m bounds for drinkable and agricultural purposes respectively.

Article 138 – The bound of spring and qanat is 500 and 250 Gaz[4] from any sides in loose and hard soils respectively. However, if the measures mentioned in this article and the one before (article 137) are not sufficient to prevent the harm, adequate extent should be added to the mentioned measures.

The article means that if the qanat is excavated in a loose soil, the bound is 500 m from each side and 250 m in case of hard soil. However, if the above-mentioned extents are not sufficient based on expert view, adequate measures can be added to the bound to guarantee that there will not be any harm to the qanat.

Article 139 – Bound is considered as the territory of the bound owner and ownership and seizure of the bound which contradict what is meant by bound is not right without owner's permission. Therefore, no one is permitted to dig a well or

[4] Gaz: a length measurement equals to 1 m. http://www.loghatnaameh.com/dehkhodaworddetail-5b2a393c881e4aa799874d4aff337d08-fa.html

excavate a qanat within the bound of anyone else's qanat or well, however, seizures not resulting in harms to qanat are allowable.

The above-mentioned article states that excavating wells or qanats within the bound of another qanat is inhibited; however, activities not imposing harm on the qanat are allowable within its bound.

References

Azad M (2000) Artificial recharge of aquifer by controlling Qanat water flow, Selected articles on Qanats. Yazd Regional Water Authority, Yazd

Labbaf Khaneiki M (2009) Final report of Shekh Mamudian Kahriz Renovation. International Center on Qanats & Historic Hydraulic Structures (ICQHS) under the auspices of UNESCO, unpublished

Safinejad J (2000a) Desert and mountainous Qanats in Kashan, Selected articles on Qanats. Yazd Regional Water Authority, Yazd

Safinejad J (2000b) Two storey Qanat of Ardestan, book of selected articles on Qanat. Yazd Regional Water Authorities, Yazd (in Persian)

Safinejad J, Dadras B (2000) Underground Dam of Vazvan Qanat in Isfahan. Iranian National Water Museum, Tehran

Salmasi F (1997) Application of computer modeling in the management of artificial groundwater recharge in the Plain of Ghorveh. University of Ahvaz Shahid Chamran, Ahvaz

Chapter 8
Qanat Related Structures

Abstract This chapter sheds light on the structures which are intertwined with qanat. For example water mills used to be built in qanat gallery in order to exploit water energy or water reservoirs were for storing drinking water. Also ice house was meant to produce ice out of qanat water and store it for summer use. This chapter shows the fact that qanat's application is not only limited to water supply, but it may encompass the other fields like generating energy, sanitation, supplying potable water, producing ice, etc.

Keywords Irrigational purpose • Drinking water • Sanitation • Water distribution • Water transfer

8.1 Watermills

A watermill is a structure constructed to grind grain by using the energy of water flow. Its main parts are a water house, two millstones, rotor blades and an axis which connects the rotor blades and upper millstone vertically. The operation of the watermill is based on the potential energy of water due to the height of water house: The deeper the water house, the more energy is generated. Sometimes the depth of a water house reaches 10 m below the depth of the qanat, in order to increase the water pressure (Fig. 8.1).

In fact water house is a cylindrical reservoir dug in the ground that receives and accumulates water. When the qanat water reaches the water house, it fills up the water house and then gushes out from a tiny nozzle at the bottom and hits the rotor blades, making the blades rotate, and imparting energy to the rotor which then turns the upper millstone. The lower millstone is motionless. The friction between the upper and lower millstones grinds the wheat into flour. In some places, several watermills were operated by the water of only one qanat. In the past in many arid regions of the world which lacked surface streams the flour was provided only by watermills operated by qanat water.

A.A. Semsar Yazdi, M. Labbaf Khaneiki, *Qanat Knowledge*,
DOI 10.1007/978-94-024-0957-4_8

Fig. 8.1 Watermill

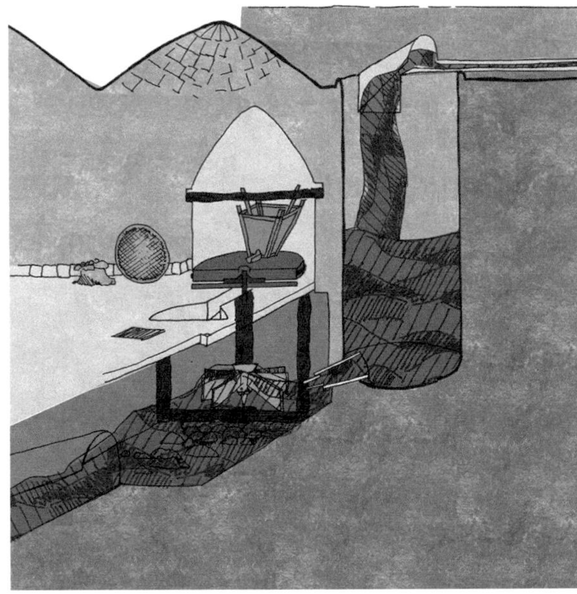

8.2 Payab

A "*payab*" is a sloping gallery connecting the ground surface to the qanat gallery before water reaches the qanat exit in order to provide people with an access to the water flow. This gallery has steps to make it possible for people to reach the water flowing in the qanat. The deeper the qanat the longer the payab. The slope of the payab is calculated so that the end of gallery meets the bottom of one of the qanat's shaft wells. The light was provided by the mentioned shaft. The size of a Payab stairway was such that at least two persons could go up and down side by side easily and their heads would not touch its ceiling. The payab was built perpendicular to the direction of the qanat gallery in order to prevent the probable collapse of the gallery.

Inside the payab, the temperature is about 15–25 °C cooler than that of outside in summer due to its underground location and proximity to qanat water (Fig. 8.2).

Some payabs were built for public use, near mosques, roads, bazars and caravanserais. But in some dry cities in central parts of Iran, many families had a private payab. The designers divided the qanat's main branch into several side branches inside the city. Each side branch crossed part of the city and then at the other side of the city, the side branches were joined to each other again. The houses neighboring the side branches had a private payab in which the owners rested during the hot days of summer and stored their food stuff. So, these houses were usually more valuable than other houses in the city. Also, there was a public payab in each part of the town for those who did not have access to a private payab, to use the water for sanitation. The payab structure was not complicated and the main part was like a room, which had a square or an octagonal plan with the following parts:

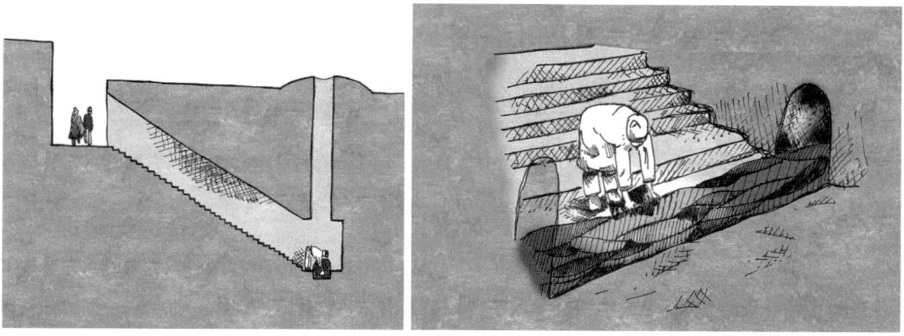

Fig. 8.2 Payab

Fig. 8.3 Basket hanging from payab ceiling to keep food stuff

- a pool at the bottom of the payab: This round or polyhedral pool had some openings through which the qanat water entered or exited.
- Some platforms: which were constructed in the walls for people to sit on. There were more platforms in a public Payab than a private one. In the private Payabs, there were some shelves to put foodstuffs on. Also, there was a rope hanging from the ceiling above the pool, tied to a basket at the end to place some food, such as meat and fruit, to be kept fresh. The payab's roof was arched, resistant to collapse (Fig. 8.3).

In the Iranian desert towns, payab transcends a mere engineering concept, but it carries many social and cultural functions. In fact social stratification is manifest in the geographical distribution of payab throughout the town, given that the higher social classes have a stronger chance to reside in the vicinity of qanat where they easily can access safe water. Therefore as qanat meanders throughout the city, its direction dictates the position of homes according to the social ranking of their owners. People did not trust the water of payab for drinking especially during day time because of free access to this water which could raise the risk of contamination. Therefore payab was usually used for sanitation purposes like laundry or bathing, and safe potable water was supplied by water reservoir which is described under the following topic.

8.3 Water Reservoirs

The water reservoir is an underground structure, constructed to store freshwater for domestic use. The reservoir was fed from a nearby shallow qanat. All of the water reservoirs had a storage tank whose dimensions depended on the amount of qanat discharge and the demand for water. Most of the storage tanks were made of local materials like brick and stone and then insulated by "*sarooj*".[1] The different parts of a water reservoir were the storage tank, the roof of the storage tank, wind tower, stairway and ornamental portal.

The storage tank is of variable dimensions and its plan is usually a circle. The whole body of the water reservoir, or the most parts of it is constructed under the ground.

There were one or more faucets in the reservoir wall at 0.5–1.5 m from the bottom. These faucets were used to transfer water from the storage tank to the "*pa shir*"[2] area, where people took the water through these faucets. The reasons why the faucets were set above the base are: to fill people's containers with water easily; to prevent the impurities and sediments deposited at the bottom of the storage tank, getting out of the faucets; and to have a high discharge because of the hydrostatic head in the tank. The water reservoirs were usually covered in order to prevent dust and also evaporation.

The water storage tank's roof is mostly dome-shaped or conic. The conic roofs are two types; flat or step wise.

The wind tower or "*badgir*" is one of the traditional structures, located in the water reservoir compound. Badgirs or wind towers cause the air to circulate in the water storage area and blow on the water surface, evaporating the water and making it cool. This process sometimes leads to a temperature difference of 30° between outside and inside the water reservoir. The number of badgirs in a water reservoir ranges from 1 to 6.

One had to go down a stairway to reach a landing named "*pa shir*" to tap the water. The number of steps depended on water reservoir's depth. The pa shir was constructed at the same level as the bottom of the water storage tank, or a little lower. There was a drain at the bottom of the pashir to discharge the waste water to a canal named "*rah ab*" and this water was then directed to a nearby qanat. Sometimes the water in the reservoir is polluted and can no longer be used, so they have to dispose of it and fill the reservoir again. Once a year the reservoir is emptied and cleaned and re-filled.

---

[1] Sarooj is a combination of lime, clay, and chipped straw, for cementing bricks or stones together.

[2] Pa shir is an area to tap water

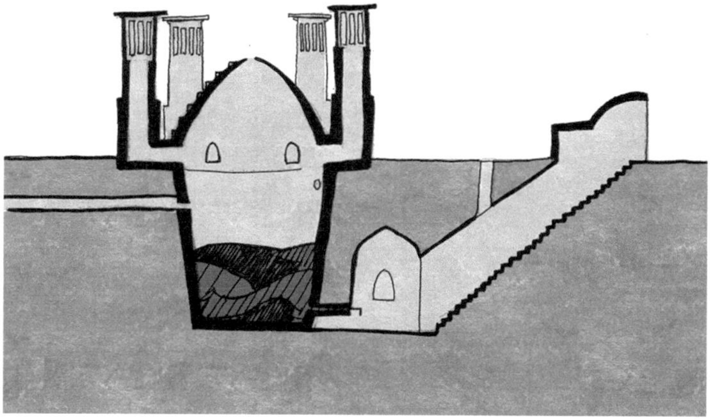

Fig. 8.4 A typical water reservoir

The ornamental portal of the water reservoir usually shows the wisdom and talent of Iranian architecture. We can see varieties of inscriptions, decorative masonry, etc. on these portals.

Water reservoirs were filled in the wet season with the water of a nearby shallow qanat and in the dry season could provide people with cool and fresh water. In order to prevent sedimentation of the storage tank, a small pool was constructed near the water reservoir and a canal conveyed the qanat's water to the pool. Then after the sediment settled, the clean water in the pool was directed to the storage tank of water reservoir. Also, salt stones were used to disinfect water by putting them in water while recharging the reservoir. The salt stones were gradually dissolved in water, producing the gas of chlorine which can kill the pathogens in water. Therefore this water was suitable for drinking and cooking, and it was carried in the potteries to the homes for such purposes (Fig. 8.4).

8.4 Irrigation Pool

If the flow rate or discharge of the qanat is sufficient, there is no need to store its water, and it is transferred to the land directly. But if the discharge is not sufficient, it is then necessary to build a pool near the exit of the qanat in order to store the qanat water. The water in the pool is then transferred to the farmers' lands for a shorter time, but at a higher flow rate than from the qanat. If there is no pool, there is no possibility for people to use the water to irrigate their lands. Otherwise, the water cannot be transferred to the land efficiently and completely because of a high loss along the way to the cultivated lands.

A pool has eight principal functions as it:

1. Stores the qanat water.
2. Increases the water flow.

3. Facilitates the system of water division and distribution.
4. Shortens the time of irrigation.
5. Reduces the waste of water.
6. Acts as a resting pool to make the impurities settle.
7. Acts as a place for social encounters because it is the potable water collecting place and the starting point for water distribution among farming units.
8. Strengthens cooperation and convergence between the farmers since it needs cleaning and maintenance annually (Figs. 8.5 and 8.6).

Fig. 8.5 Irrigation pool

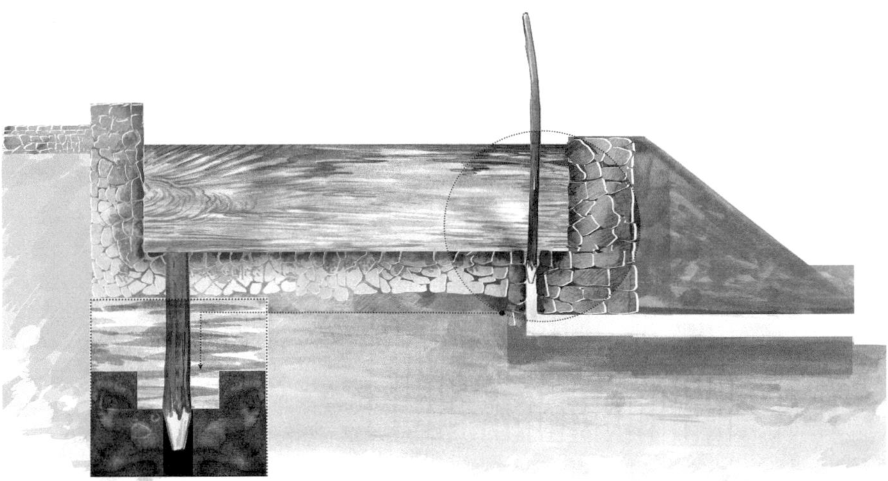

Fig. 8.6 A stick put into the pool drain to store water

8.5 Ice-House

The freezing nights during winter in the desert made it possible to build ice-houses to produce and keep ice for the warmer seasons. Ice-house is related to qanat, for the water needed for producing ice comes from a nearby qanat. On the outskirt of almost every desert town, one can see some weird conic structures with two walls that protrude from the both sides of the structure in opposite directions. The walls are oriented east west to be able to shade some rectangular pools just behind the wall, each 10 m long and 50 cm deep. During the freezing nights of winter, the water of qanat was directed into the pools where it could freeze and form a layer of ice. The next night more water was poured into the pools to form another layer of ice on the top of the layer of last night. This way a thick layered mass of ice could be produced, which had to be broken into smaller pieces and brought to the ice house reservoir. The pieces of ice were stacked and then covered by straw every other one layer by layer until the reservoir became full. The layers of straw could insulate the layers of ice and help them last at least until the mid summer. An ice house reservoir is 10–15 m deep and 10–20 m in diameter, dug into the earth, covered by a high dome. Its overall structure is similar to that of a water reservoir except it lacks wind towers. Also a stairway was devised in one side of the reservoir, through which one could access the pieces of ice. A typical ice house was built by stone, brick and lime and the inner side of the reservoir was lined by a coat of "Sarooj", a traditional mortar made up of lime, clay, cane fiber and ash. The shading walls were built by brick and plastered with clay. The walls are 1.5–2 m at the base and get thinner at the top. The walls are 15 m high (Dehghani 2009: 74) (Fig. 8.7).

Generally affluent people were the costumers of ice houses in the past, because ice was not considered a necessary commodity and normal people could live without it. Given that the high social classes used to dwell in the cities the ice house were mostly built around the cities. The ice houses increased in number when the economy thrived, so the number and distribution of the ice houses herald the economic situation of the Iranian cities in the course of history. Therefore the capital accumulation could give rise to a boom in the number of ice houses, and the capital accumulation itself was rooted in a change in the production means. The other way around is also valid. Thus any recession could lead to a decrease in the demand for ice and accordingly the number of ice houses diminished (Bahadori Nejad and Dehghani 2010: 96).

8.6 Chah-khane; Well House

In the past, public baths or houses belonging to the affluent people in the vicinity of a qanat used to have a well which went down to the qanat gallery directly or ended up a few meters away from the main gallery and then reached the water flow through a short side tunnel. These wells were wider than the normal qanat shaft wells, some

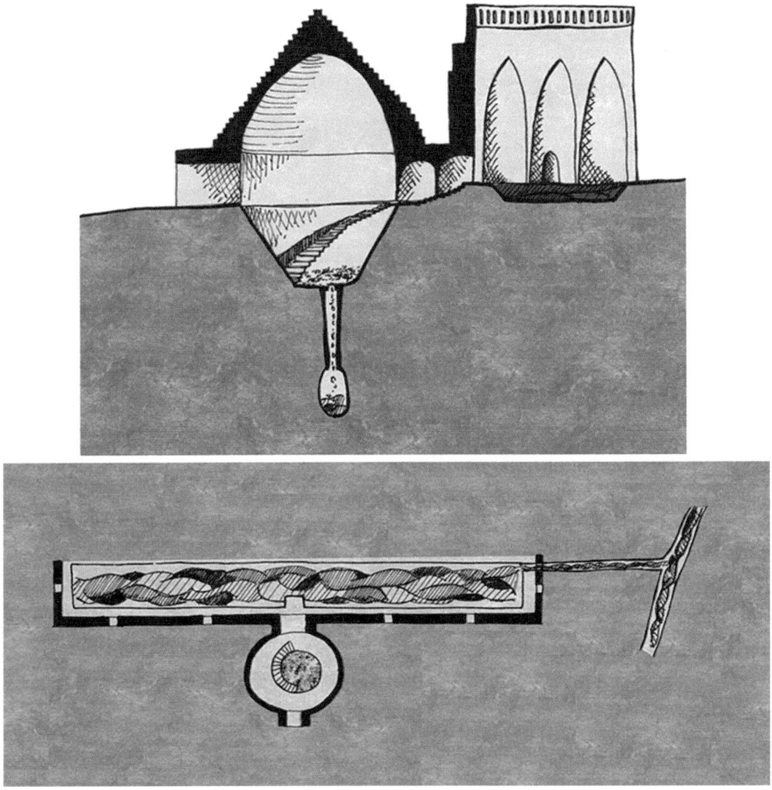

Fig. 8.7 Profile and plan of a typical ice house

1–3 m in diameter. A Chah-khane was located in a chamber, consisting of a well, a pool, a reservoir, a seat for the operator, windlass and bucket. There were three types of Chah-khane as follows:

Type 1: Chah-khane for large public baths: This type was built near the public baths that enjoyed big water reservoirs and were used by people round the clock. This type occupied a considerable area. Water was pulled up by big buckets through a pulley hanging above the well, being operated by the force of such animals as donkey, cow or camel. A sloped passage was also built close to the Chah-khane, through which the animal could walk down and pull the rope behind them. When the animal reached the end of the passage, the bucket appeared and poured its water out, then the animal turned around and walked up the passage and sent the empty bucket down the well. This water went to a reservoir through a ceramic pipe locally called "Tanboosheh" (Fig. 8.8).

Type 2: Chah-khane for small public baths: in the small public bath where water was needed 24 h a day, water was extracted by a hand operated windlass. A windlass was installed above the well and a bench was devised there so that a worker could

Fig. 8.8 Chah-khane for large public baths

sit on it and pull water whenever needed. The water was poured into a small pool nearby from which water was directed to a reservoir and then to the bath pool.

Type 3: Domestic Chah-kane: in the past, the houses built on the qanat direction had a chance to have a private access well to the qanat water. These wells were dug to get access to the qanat current, and the well opening was located in the second storey or even the roof of the same house. Near the well opening, a small pool with a capacity of a few cubic meters of water was built which stored the water pulled up by a windlass. A faucet was devised just beneath the pool, that allowed the water down into a pipe leading to kitchen sink and yard pool. This water sufficed to meet their need for some 10–20 days, and afterward the roof pool should be re-filled (Semsar Yazdi 2014: 50) (Fig. 8.9).

8.7 Bookan

In the absence of vehicles and means of transportation in the past, qanat practitioners and workers had to stay away from their houses, near the qanats under construction. So, they had to build temporary houses for themselves in those places. Those temporary houses were named "Bookan". "Bookan" was an area, constructed near the qanat. Its roof was at the same level as that of ground surface, and workers could enter that through some stairs. The thickness of Bookan's roof was about 2 m and the workers built it together under the ground. Nowadays, "Bookans" like Payabs have lost their functions, and today there is no active "Bookan" left. But in the past, existence of "Bookan" was essential, especially in the long qanats; and

Fig. 8.9 a typical Chah-khane in a private house

without it, the work could not be continued. The length of a plain qanat may reach to 30 km and more, so, there might be a long distance between workers' house and their work place. It was impossible for the workers to travel this distance and back every day, so construction of "Bookan" was inevitable for qanat workers. Moreover, the "Bookan" was used as a small forge for sharpening the pick axes. The workers, who were resting in the "Bookan", prepared food for other workers who were digging qanat and also to sharpen the pick axes for them. The forge in the "Bookan" was connected to the outer space through a chimney.

The workers lit the forge and sharpened the dull pick axes on it by pounding with a sledge hammer on them. This process was named "Kolang Keshi". The method of building a "Bookan" was as follows: at first, the workers dug a slant gallery in the ground with appropriate slope, and at the end of that gallery they built an underground room in a conical shape. They were cautious about the probable cave in and so they were very careful in constructing a "Bookan".

In order to prevent a collapse, they used to leave some columns of earth intact in the middle of the room. They placed the dirt taken out around the entering hole to prevent rain from coming in. Another hole was dug to let the smoke of cooking out. Sometimes this hole served as a chimney and sometimes as an air vent, because the air for breathing was also provided by this hole. The size of the "Bookan" depended on the number of the people who were to stay there, the more the people, the bigger the "Bookan".

Collapse is likely due to the instability of earth after abandoning the "Bookan", unless the "Bookan" has been dug in a hard ground.

Depending on the number of muqanis, several niches were dug into walls, 0.5 m above the floor. These holes were called small chambers or sleeping platform. After digging the main parts of the "Bookan", they also gouge some smaller niches in the wall of living room, so that the muqanis could put their personal stuff in it; there was also a special place for putting the fat lamps and torches in it, and another place for putting the instruments used for digging. A place separated from the living room was also designed for cooking. Almost every 3 km along the gallery, a new "Bookan" was built; and if the gallery was dug in a hard area, the muqanis had to stay in one "Bookan" for a long time (Semsar Yazdi and Labbaf Khaneiki 2010) (Fig. 8.10).

8.8 Maqsam

In the case of high discharge qanats shared between several villages or farmlands, the water flow may be divided into separate streams before being utilized. There is usually a special structure named maqsam –water distributer- which consists of some outlets of different sizes on which the shareholders have agreed.

If two or more villages or agricultural areas are entitled to a particular qanat, a maqsam is built across the canal immediately after the qanat water reaches the surface. In fact a maqsam is a small dam with two or several outlets either of which directs water to a particular area. By means of a maqsam, it is possible to distribute water among the areas. Maqsam is usually constructed with wooden frames or stone and a mortar of lime and clay in the past, but substituted by concrete today. In some regions, maqsam is built inside a chamber whose door is always locked and only mirab who is in charge of water division has the right to enter the chamber, because of fear that someone illegally manipulates the outlets and changes the water shares (Fig. 8.11).

Fig. 8.10 Bookan

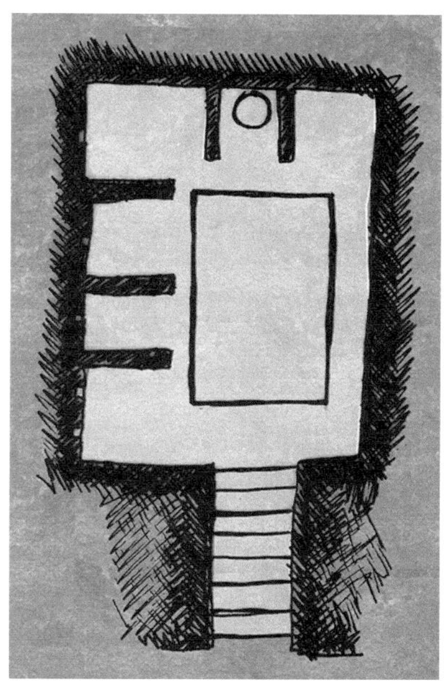

Fig. 8.11 Maqsam for dividing water

8.9 Aqueduct

Aqueduct spans a gully or valley in order to convey qanat water from one side to the other. This structure is akin to Roman aqueduct in its shape and function. Aqueduct is built somewhere in the gully whose soil structure can support the weight of the columns, and it is made of stone, brick and lime mortar. The size of the aqueduct is engineered such that it does not give way under the fury of seasonal runoffs. Unlike the Roman aqueducts which transfer water in a covered channel, this aqueduct enjoys an open ditch atop the structure, in which qanat water flows across the valley. A good example of such Aqueducts is located in Kharanaq in Yazd province, which is still in active use (Fig. 8.12).

Fig. 8.12 Qanat aqueduct

Glossary

Ab Anbar Water reservoir which is an underground structure, constructed to store freshwater for domestic use. The reservoir was fed from a nearby shallow qanat. All of the water reservoirs had a storage tank whose dimensions depended on the amount of qanat discharge and the demand for water. Most of the storage tanks were made of "sarooj" a combination of lime, clay, and chipped straw. The different parts of a water reservoir were the storage tank, the roof of the storage tank, wind tower, stairway and ornamental portal.

Ab Andaz Ab Andaz is a lateral tunnel that diverts water from the watermill. In case the watermill is out of order or under repair, Ab Andaz is used to convey qanat water.

Ab Gaz Sometimes the qanat workers have to stay and work in water for many hours a day, which gives rise to a kind of skin disease locally called Ab Gaz which literally means "stung by water". This disease is known by skin rash and irritation which can be cured by an ointment made of pomegranate skin soaked in water.

Ab Harzi Ab Harzi means flow easing. It is enough to ease the flow in the qanats which are not crumbling. This practice is called "Ab-Harzi" which involves walking along the tunnel from the qanat exit to the mother well in order to control the pace of water flow. This practice may take the workers several days, depending on the qanat length.

Ab Pishbari If the tunnel cuts through water saturated area, water infiltration is much enough to pave its own way toward the exit point, otherwise the workers have to do something to ease the water flow which is called "Ab Pishbari". The tunnel slope is so gentle that water needs to be moved forward by a shovel. Thus a qanat master follows the water current along the tunnel and explores the tunnel all the way to remove any possible barriers on the way of water.

Abgan Refer to Tare Kar

Araj Araj is a traditional unit of length especially used in qanat measurements. Araj equals the distance between the middle finger's tip and elbow some 50 cm. Araj also means a tunnel with a width of 60 cm all the way with no bulge or dent on the walls. Any slight unevenness on the walls of tunnel can hamper air circulation, resulting in stuffiness of the tunnel. According to the qanat masters, the dimensions and shape of a gallery can play an important role in ventilation. The gallery should be 60 cm wide all the way where the workers are in trouble with gas.

Arghamchi Arghamchi is a cotton rope that has strong fibers, made of three or four smaller cotton ropes interwoven. This rope is used mostly in double windlasses which need stronger and thicker ropes, though its disadvantage is that it absorbs water in wet conditions and gains weight over time. Moreover this type of rope comes to rot and fall apart when wetted for a long time.

Asiyab Watermill. A watermill is a structure constructed to grind grain by using the energy of qanat water flow. Its main parts are a water house, two millstones, rotor blades and an axis which connects the rotor blades and upper millstone vertically. The operation of the watermill is based on the potential energy of water due to the height of water house: The deeper the water house, the more energy is generated. Sometimes the depth of a water house reaches 10 m below the depth of the qanat, in order to increase the water pressure.

Baghal Baghal is a traditional unit of which equals 175 cm.

Baghal-Bor By-pass. Where the tunnel hits a boulder, the workers bypass the boulder and then get back to the general direction of the gallery. This practice is called "Baghal-Bor".

Band-e zirzamini Underground dams which are possible to build in the tunnel wherever the soil structure is sturdy and is not vulnerable to water build-up, otherwise it collapses. Thus the qanat gets blocked every year in autumn until late winter in order to save the qanat water flow.

Bandesh Blockage. A mass of spoil and debris which obstruct the qanat tunnel.

Bandsar Refers to Goorab.

Benow In Bam region sometimes the workers come across temporary springs while digging the tunnel, which are called "Benow". Benow is actually small reserves of water trapped in underground cavities, which spring up in the tunnel and flow down.

Bil-e qanat Type of shovel for digging qanat. This type of shovel is smaller than a normal one, with a shorter handle which can move about in such a small place underground. This shovel is used to gather the spoil and dump it in the buckets in order to lift it out. A typical qanat shovel enjoys a 15 cm handle in length.

Bizesh Bizesh is a type of Devil. In some cases the worker can pinpoint the Devil spot accurately and even starts digging overhead but he may come across an impassable obstacle like a boulder. Now it is very difficult to dig through such a boulder from down to up, so the worker prefers to do the other way around in order to have full control over the digging. In this way he can strike his pickaxe more strongly. Therefore he digs a diagonal tunnel from the main gallery roof toward the well, just over the boulder. This diagonal tunnel is named Bizesh which drains out water at the well, allowing the worker to dig the rest of well down to the gallery.

Bookan The qanat workers usually preferred to stay in the place of their work, due to the long distance they had to travel to their homes and lack of transportation. They used to get back home once every two month and they had to stay in an underground hut called "Bookan" all over this time. A typical Bookan enjoyed a room for sleeping, a chamber for cooking and a place for repairing and sharpening the pick-axes.

Chah-e Gowri Refer to Chah-e Kermani

Chah-e Kermani According to the qanat masters, in distant past the qanat practitioners did not have sufficient expertise to dig circular wells with observing Passa, so they resorted to digging rectangular wells instead. The locals call such rectangular wells "Kermani". They believe these wells have been dug by Zoroastrians long ago before the advent of Islam. The dimensions of such wells are usually 80 by 120 cm, which can be found across Iran.

Chapar In crumbling wells an instrument named "Chapar" is used to protect the worker against the objects falling down, while working at the well bottom. Chapar is a semicircular wooden umbrella rotating around its stand and covering half of the well cross section all the time. Its stand is some 1.5 m high, which rests on the well bottom, providing a safe place against the soil and stones falling down.

Char Kaval When the tunnel reaches a shaft well, the workers put four hoops across from each other just at the bottom of the well. They call such wells Char-Kaval which means a well with four hoops at its bottom.

Charkh Kar A working team which consists of at least four workers closely connected like the cogs of a machine. The team head is the qanat master (moqqani) who is more experienced than the other workers and digs the tunnel end into the soil. A worker sits behind the master and collects the excavated soil and puts it in the buckets which are carried by another worker along the tunnel and hooked to a rope hanging from the nearest well. Finally a worker operates a windlass on the ground to coil up the rope and pull up the debris bucket and dump it around the well opening. These workers perfectly team up to make the most of their energy and time.

Charkh Kesh Charkh Kesh works with windlass. He operates the windlass with his hands or feet to send the rope down the well and pull things and people up.

Charkh-e Chah Windlass. Type of pulley by which workers can lift out the spoil from a qanat well. A windlass is composed of four timbers crossing each other two by two, and four poles go through the crosses at right angle to attach them together and form a pulley. The two crosses are placed parallel about 1 m away and then nailed and tied to the four poles to get them firmly attached. A stronger pole goes through the middle of the pulley to act as an axis around which the pulley rotates. This axis is placed on two wooden stands which are firmly fixed into the ground just on two opposite sides of the well opening.

Cheragh Karbit Carbide lamp. This lamp was introduced in early 1970s and widely used by qanat workers across Iran. Even today this lamp could have retained its position despite availability of electric lamps. Carbide lamp is based on a chemical reaction between calcium carbide (CaC2) and water which produces acetylene (C2H2). This lamp consists of two small containers one on top of the other. The upper container is filled with water which drips out of a tiny hole at the bottom. Some calcium carbide is put in the lower container where it reacts with the water and produces acetylene gas. This gas gets out through a small pipe attached to the lower container, and burns with a bright white flame. When all of the carbide has been reacted, in the lower container a wet paste of slaked lime (calcium hydroxide) is left.

Cheragh Mushi Cheragh Mushi is a lamp that works with petroleum.

Cheragh sooz End of the qanat gallery. In some cases the gallery advances into the aquifer a few meters up slope from the mother well. This part of gallery is called Cheragh sooz. A "Cheragh Sooz" lacks shaft wells all the way, and is dug just to increase water infiltration into the gallery in groundwater saturated area.

Chilar Mass of pebbles which may be found underground and their size varies from a pea to a walnut. In the water production section these pebbles benefit the qanat.

Choobe gheyb yab Diving rod to find water underground. A practitioner whittles a Y-shaped rod off a nut tree and hold it loosely in their hands while walking gently. They believe that the presence of groundwater can affect the rod by making a feeble impulse which can be sensed under one's fingers.

Dalv Gir Dalv Gir takes the buckets off the hook on the surface and dumps the debris around the well mouth.

Dalve Bucket which is used to lift out the spoil and haul it onto the earth surface. In the past the buckets were made of tanned sheep skin, but today rubber is used to make them. In the past the workers sewed a tanned sheep skin round with some leather strings cut out of the same skin. Afterward the rim of the bucket was folded back and sewed so that a flexible thin branch of pomegranate can drive into it all the way round. This way the bucket remained open, ready for use.

Dam Dam implies oxygen shortage which takes place in the wake of air stagnancy, for the oxygen molecules would be absorbed by the mineral materials in the closed space. According to the qanat masters, "Dam" puts out a light or a candle where it builds up in the tunnel. Before the workers go down a well, they tie a light to a rope and send it down the well to see if the light keeps burning or goes out to make sure that there is enough oxygen down there.

Dam dar Choky, stuffy place. A place with Dam. Refer to Dam.

Dang Subdivision of an irrigation time unit which is one sixth of a standard unit. To measure a Dang, some marks are cut into the inner side of the small bowl which divides the certain unit of time into the shorter fragments (refer to Saat-e Abi).

Dast Charkh Refer to Charkh Kar.

Dastak Dastak is a short tunnel deriving from the side of main gallery. A Dastak rarely exceeds a few meters in length, and it may be dug for increasing the area of water seepage into the gallery.

Dastangineh In some intersecting qanats at different levels, the workers devise an outlet through which the qanats can be connected when needed. For example if the upper qanat is not in good shape and needs repair, they divert its water to the lower qanat for more convenience while working. In case the lower qanat is much deeper than the one above, its gallery gets connected to the nearest well of the lower qanat by a short horizontal side tunnel which is called Dastangineh.

Devil In the saturated area it is not possible to dig the wells deeper than the water table, because water level comes up and drowns the worker. Therefore the workers give up digging a well as soon as it reaches the water table and turn to digging next well in the hope that groundwater would be drained out by the qanat gallery in the mean time and water table goes down. If no considerable drawdown occurs in water table, the workers manage to dig the rest of the well from the tunnel upward. This practice is called Devil. Water pours down and flows away as the worker digs overhead. "Devil" is rectangular in shape with a dimension of 80 by 120 cm.

Do Sar Do Sar or double is a type of windlass which enjoys two ropes with two buckets working parallel. In deep wells the workers prefer to use Do Sar in which one bucket goes down the well while the other is coming up. The weight of the empty bucket going down helps the worker pull up the full bucket coming up.

Dogali When a rope is not long enough, the qanat workers tie it to another rope with a special knot named "Dogali". They twist the two ropes and knot them such that they attach more strongly as they are pulled apart.

Dool Refer to Dalve.

Estakhr Irrigation pool. If the flow rate or discharge of the qanat is sufficient, there is no need to store its water, and it is transferred to the land directly. But if the discharge is not sufficient, it is then necessary to build a pool near the exit of the qanat in order to store the qanat water. The water in the pool is then transferred to the farmers' lands for a shorter time, but at a higher flow rate than from the qanat. If there is no pool, there is no possibility for people to use the water to irrigate their lands. Otherwise, the water cannot be transferred to the land efficiently and completely because of a high loss along the way to the cultivated lands.

Fenjan Refers to Saat-e Abi.

Fetileh To make Sazoo, the date palm fibers are woven together to form some thin strings which are called "Fetileh" and then each two strings are twined to form a primary rope which is called "Menal". Afterward they use a special weaving machine to twine the Menal and turn them into Sazoo.

Gamaneh Gamaneh is a test well which is sunk first of all in order to confirm the existence of groundwater and to assess its depth. After finding a suitable place for building a qanat through natural indications, the first step is to sink a test well. In many cases more than one test well are dug in order to get a better view on the situation of groundwater and make sure about the feasibility of qanat construction. The optimum depth of test well is associated with the geological and topographical conditions of the region.

Gaz Gas. Gaz is poisonous gas which is continuously being re-produced in the ground so that one can see its tiny bubbles coming up in water. Gaz is more common in mountainous qanats and becomes more plentiful during cloudy days. Presence of harmful gas can pose a serious threat to the workers and even bring their work to a standstill. According to the qanat workers, it is hard for them to breathe in Gaz though their lamp does not go out.

Gaz Wedge. Long stones were embedded 40–50 cm into the wall of a well, wedged firmly in place with a piece of wood locally called "Gaz" in order to a make a base for blocking the well in the middle.

Gelband Gelband dumps the spoil in the buckets and hands it over to the Lasheh Kesh.

Gel-e Morde The mud found in a well that has collapsed is called "Gele Morde" which means "dead mud". It is very difficult to empty such mud since new mud replaces and rushes back as the workers empty it.

Gom Refer to Kaval.

Gonbat Qanat masters believe that the shaft wells in which air circulation takes place from top to bottom are more subject to collapse. In this case the well begins to collapse mostly in the middle, forming the shape of a dome. The workers call such wells Gonbat.

Gonbate Zar Refer to Shoolat.

Goor Sometimes the workers encounter smaller boulders which can be moved or rolled. In this case they dig a hole just by the boulder and roll the boulder down the hole by means of a lever. The hole is dug on left or right of the boulder to bury it and get it out of their way. This hole is called "Goor" which means grave in Persian.

Goorab Goorab is a traditional dam built for harvesting seasonal runoffs. During the rainy seasons, the runoffs flowing down the surrounding hills were directed to the dams. In the dams water built up and gradually percolated the earth into the aquifer which fed the surrounding qanats.

Gor Work shift. Two qanat masters took turns working in the gallery once every 6 h called a Gor. A qanat master worked in the tunnel from sunrise to noon while the other was busy gathering firewood and preparing meal for the workers. At noon the workers took a break and had their lunch and then began the second shift of their work.

Gordeh Gavi A soft clay formation in which the tunnel should be shored up with hoops otherwise a sudden collapse is inevitable. This formation is called Gordeh Gavi which literally means "cow's flank", because pieces of mud the size a cow's flank may fall down and destroy the tunnel.

Goriz Gah Goriz Gah refers to hole dug into the tunnel side in which the worker can take shelter where he manages to remove an obstruction behind of which there is a water build-up. The worker escapes to Goriz Gah as soon as the obstruction starts to give way.

Gozareh The short tunnel or well which connects two intersecting qanats is called "Gozareh" which remains closed in normal situations. In some intersecting qanats, the workers devise an outlet through which the qanats can be connected when needed. For example if the upper qanat is not in good shape and needs repair, they divert its water to the other qanat for more convenience while working.

Hadd-e taran va khoshkan Refer to Marze Taroon Khoshkoon

Halghe mile To get started on digging a well, first the worker puts his pick axe on the ground surface the way the pick axe can be rotated around its point and makes a circle on the soil with the end of its handle, like compasses. The handle is usually 35 cm long, so the circle would be some 70 cm in diameter. Afterward the worker starts digging inside this circle which called halghe mile.

Haranj Refer to Kush.

Harim Harim means the bound of qanat and mostly pertains to the hydraulic vicinity of qanat in which any water extraction may have negative impact on the qanat discharge. The bound of qanat depends on a variety of factors including the surrounding geological formations, type of aquifer, proximity of other groundwater sources, depth of gallery, etc.

Havabin Shallow mountainous qanats with fluctuating discharge are called havabin which literally means looking up at the sky. If a test well reaches a shallow groundwater, such a qanat would be going to have an unreliable fluctuating discharge.

Havakesh Havakesh is the most common method to curb the poisonous gas by building a ventilating duct. This duct can be built only in the tunnels which have been lined with ceramic or concrete hoops, because the 20 cm space between the tunnel roof and the hoops serves to move the stagnant air in the tunnel.

Jofte Badoo Twin wells. Usually the first well dug immediately after the test well is called Jofte Badoo which locates in water transport section. Jofte Badoo consists of two wells dug some 1.5 or 2 m away from each other to better ventilate the wells especially where soil emits harmful gas. Some of the geological formation gives off poisonous gases that are obviously harmful to the workers, so the workers dig two wells simultaneously and connect them every 10 m by a short horizontal tunnel in order to ease air circulation as they go down. In this case fresh air comes down from a well and goes up through the other.

Ju Ditch, brook. Water current on the earth surface after the qanat exit point.

Kabar Refer to Kaval.

Kaf shekani Tunnel deepening. Kaf shekani means the qanat gallery being deepened from its exit point to the mother well all the way in order to access deeper layers of saturated area. The floor of qanat tunnel is dug deeper as water table goes down. Deepening of tunnel results in more water seepage into the tunnel and sometimes the amount of water is so much that the workers have to widen the bottom of tunnel in order to better ease the water flow.

Kalleh Qoochi Kalleh Qoochi is a type of roof in a qanat tunnel, which is built of two flat stones braced against each other in the shape of a pitch roof. This type of roof is used to shore up the tunnel only with stones.

Kamargir A lid installed in the middle of the well to block it. The workers embed a row of bricks in the well side all the way round, when they want to build a base for a lid in the middle.

Kamvar The workers pile the excavated materials around the well mouth to build a round dyke which can ward off the flood. These round dykes are called "Kamvar" which make up a landscape in an Iranian desert, which looks from air like a land overrun with rows of anthills.

Kanvar Refer to Kamvar.

Kanak Sometimes a part of tunnel collapses and leaves an obstruction, making water build up in the tunnel. In this case water level goes up and makes the worker mix up this accident with a reverse slope. The worker may be misled by a collapse-induced obstruction and accordingly he may decide to build the tunnel steeper. In this case such a steep slope may play havoc with the structure of the tunnel. To prevent such a mistake, the worker digs a small hole into the tunnel side just after he finishes his daily work. The hole is some 10–15 cm deep just above the water level in the tunnel. This hole is called "Kanak" which serves as an alarm if there would be any obstruction in the tunnel. The day after, if water reaches the hole, it means that a collapse has obstructed somewhere in the tunnel, making water build up and go up to that point.

Kaval Ceramic or concrete hoop to shore up the qanat gallery. Kaval is oval in shape, formerly made of clay but now of concrete to prevent a tunnel from collapsing. Shape of Kaval is such that it can resist more pressure on its top in the tunnel. The workers manage to install oval hoops as soon as the tunnel cuts through soft crumbling soils. In case exposure to air does not step up the collapsing of soil, the workers put off the lining until the tunnel reaches the next well; otherwise they install the hoops one after another as they advance toward the next well.

Kaval Sargardani If a part of tunnel is lined with reinforcing hoops (Kaval), the hoops are taken from the old level and re-installed in the new one concurrently with the deepening of tunnel. This practice is called Kaval Sargardani which literally means turning the hoops.

Khak-e Solb Hard soil with low permeability.

Khar shotor Alhagi, a desert wild plant which is used to locate a new qanat. A concentration of Alhagi indicates that there is groundwater at a depth of less than 15 m, because the roots of Alhagi feed on fresh water at such a depth or less.

Khoshke Kar Each qanats consists of two different parts; water production section and water transport section locally called Khoshke Kar. Water transport section cuts through a dry soil and just serves to convey water to the earth surface.

Khoshkoon Refer to Khoshke Kar.

Kolang Andaz Kolang Andaz refers to the bound of qanat well which is considered a circle around the well mouth between 2 and 4 m in diameter. To specify the bound of well, a worker sits down in the well while digging it, after the well reaches the depth of some one meter. Afterward the worker musters all his force and throws his pix-axe back over his shoulders. Wherever the pick-axe lands it is considered the far end of the well vicinity which should be observed by all people.

Kolang Dar Qanat master. Kolang Dar digs the tunnel forward and he heads the qanat working team. He should be the most skillful worker who knows how to connect a well to another from the bottom by digging a tunnel as well as many other engineering details. Other workers can become a qanat master one day if they try to learn all the knowledge and skill required.

Kolang-e Abgir Kani A special pick to dig water production section, whose point is about 70 cm long. In case water level would be below the worker's knee, it turns out that somewhere in the tunnel has a reverse slope not letting water flow down freely. So the worker looks for this reverse slope and fixes it with the help of this pick axe.

Kolang-e Barg-e Bidi Kolang-e Barg-e Bidi is a willow leaf like pickaxe for digging in soft soils.

Kolang-e Chah Kani Kolang-e Chah Kani or Well digging pickaxe has a relatively short handle which is some 30 cm long and weighs 3 kg. This pickaxe is short handled because it is used for digging wells which are not more than 80 cm in diameter and a longer handle is not easy to wield there.

Kolang-e Dandane dar Kolang-e Dandane dar is a ragged pickaxe for digging in mud.

Kolang-e Kar Kani Kolang-e Kar Kani or tunnel digging pickaxe weighs some 2 kg with a 50 cm long handle to dig forward into the soil. When the workers want to extend the qanat tunnel, they prefer to use this type of pickaxe which is relatively light and does not tire them out.

Kolang-e Nok Gongeshki Kolang-e Nok Gongeshki is a sparrow's peak like pickaxe for digging into rocks.

Kolang-e Saghf Kari Kolang-e Saghf Kari is a roof building pickaxe for digging the tunnel roof in the shape of a pitched roof.

Kolang-e Sang Kani Kolang-e Sang Kani is a stone digging pickaxe for hard formations.

Koor A qanat that has been clogged up by the buildup of sediment.

Kooreh Refer to Rahrow.

Korom In order to dig the water transport section, on the surface the workers mark the places where next wells are to be dug one after another, by piling up some soil and sand. These piles show the direction of the new qanat on the ground surface from the test well down to the exit point. In Yazd region these piles or marks are called "Korom".

Kush Open trench. In the region of Bam, the qanat tunnel is dug as an open trench where the tunnel approaches the exit point and its depth does not exceed 4 m. This open canal is called "Kush" which gets shallower and shallower as approaches the earth surface.

Lamal Kardan To insulate the tunnel floor in the water transport section, the workers also used to tread clay mud into the cracks and gaps on the floor. This practice is called Lamal Kardan; the worker spreads some clay mud on the tunnel floor and walks on it so the clay goes into the cracks through which water may escape.

Lasheh Kesh Lasheh Kesh drags the buckets along the tunnel. He hooks the bucket of soil to the rope after dragging it all the way from the gallery end to the first well down slope. In the past this person was usually a little boy some 10 or 12 years old, whose small size allowed him to move along the narrow tunnel more easily than an adult.

Layroobi Layroobi means tunnel cleaning. Some qanats need to be cleaned every once in a while and some other do not. For example the qanats which run through clay formations should be cleaned very often but those qanats in sand and gravel are not cleaned that often.

Lisab Refer to Marze Taroon Khoshkoon.

Madar Chah Mother well which is sunk usually into an alluvial fan to a level below the groundwater table. A typical qanat consists of a series of vertical shafts in sloping ground, interconnected at the bottom by a tunnel with a gradient flatter than that of the ground. Mother well is the first and deepest well up slope, in which a large inflow of water shows the qanat is in a satisfactory state. If a qanat is extended so far that another well is needed, the new well would now be the mother well and the former one would be a normal shaft well.

Madar Irrigation cycle, water rotation. Water for irrigation is owned by shares. In fact the farmers take turns bringing qanat water to their land. For a particular shareholder, the interval between two irrigations means an irrigation cycle or rotation pattern of irrigation water called "Madar". For example, if a farmer has an irrigation right of 2 h within a 6 day irrigation cycle, it means that he has the right to water his land just for 2 h once every 6 days. The duration of the irrigation cycle ranges from 6 to 21 days in all over Iran, but the average is between 6 and 16 days in most of the country, which is associated with the cropping pattern and soil condition. In terms of wheat and barley which are the most common crops in Iran, the best interval between two irrigations is 12 days, and that is the length of the most common irrigation cycle.

Makooyi A qanat gallery with a pitched roof to prevent water drops from falling down directly on the worker's head. Such a gallery is dug into the aquifer below water table. Such ceilings not only make water drops slide down the steep planes, but also reinforce the tunnel and reduce the risk of collapse.

Manjidi Refer to Manjil.

Manjil Buckets made of rubber.

Mard Khaneh A shack built of wood and branches close to the qanat well, where the workers stay overnight while working in the same well.

Marze Taroon Khoshkoon Transition zone. The border between water production section and water transport section is called Marze Taroon Khoshkoon which is not stationary all the time and it moves to and fro with amount of precipitation and accordingly the expansion of aquifer.

Mat In the plain areas the qanat masters envision a very gentle slope for the tunnel in which water can flow only very slowly. The locals call such a slow current "Mat".

Mazhar Qanat exit point where the tunnel and the ground surface eventually intersect, where water appears.

Mehrabi Refer to Makooyi.

Menal To make Sazoo, the date palm fibers are woven together to form some thin strings which are called "Fetileh" and then each two strings are twined to form a primary rope which is called "Menal". Afterward they use a special weaving machine to twine the Menal and turn them into Sazoo.

Mile Chah Shaft wells sunk along the tunnel to connect the surface to the horizontal gallery. The main application of these wells is to lift out the spoil and excavated materials from the tunnel on to the surface. They also provide access and help ventilate the tunnel and provide more oxygen for the workers. These wells play an important role in repairing the qanat, by making it possible to send down the needed facilities and tools and remove the debris. The shaft wells cut short the time needed for qanat construction or repair, and reduce the relevant expenses. A shaft well is between 80 and 100 cm in diameter.

Mirab Professionals who are in charge of distribution of water among the farms, and are paid a certain salary by all the shareholders. Mirab should be quite familiar with the network of ditches leading water to the farms, because he has to factor in the time that water takes to travel the ditches and reach each farm. Mirab has a notebook too, including all the irrigation rights in detail, so if the shareholders want to sell or buy any right they should inform mirab of any transaction.

Moqanni Refer to Kolang Dar.

Nay Refer to Kaval.

Nimche Nimche is a type of pickaxe for digging the tunnel floor before having a water flow in it.

Pa Raf Foothold. Pa Raf means small niches dug into both sides of a well through which the workers can climb up and down the well. These niches are built some 100–125 cm away from each other, such that each niche faces the distance between the two apposite niches.

Palooneh In the water production section, the hoops should not be fitted together completely with no distance between them, because it can insulate the tunnel and reduce the water infiltration. In this case, the workers consider a small distance between the hoops, and they put some pebbles and pottery pieces behind the hoops such that the distance would be covered not to allow the clay and mud to leak in, while keeping the water seepage into the tunnel. This practice is called Palooneh.

Passa The qanat worker digs the well in a clockwise spiral way as if he climbs down a tower stairs. Therefore one side of the well bottom is some 30–40 elevated than the other, which is locally called "Passa". A skilful worker is a worker who can observe "Passa" properly.

Pavandeh Pavandeh is a noose which is tied around one stand of a windlass, to brake the windlass and keep it motionless when needed. The windlass resumes working if "Pavandeh" releases it.

Payab A payab is a sloping gallery connecting the ground surface to the qanat gallery before water reaches the qanat exit in order to provide people with an access to the water flow. This gallery has steps to make it possible for people to reach the water flowing in the qanat. The deeper the qanat the longer the payab. The slope of the payab is calculated so that the end of gallery meets the bottom of one of the qanat's shaft wells. The light was provided by the mentioned shaft. The size of a Payab stairway was such that two persons could go up and down side by side easily and their heads would not touch its ceiling. The payab was perpendicular to the direction of the qanat gallery in order to prevent the probable collapse of the gallery. Inside the payab, the temperature is about 15–25 °C cooler than that of outside in summer due to its underground location and proximity to qanat water.

Pih Sooz Pih Sooz is a ceramic pear-shaped lamp with a small handle at the back. This lamp was fueled with vegetable oils such as oil of cotton seed, castor or sunflower. The lamp had a spout on which a wick was placed with its other end soaked in the oil which went up the wick and burned with a yellow flame.

Pish Kaval Sometimes the soil from the tunnel sides and roof comes down and suddenly collapses. When the tunnel cuts through such a soil, the digging of tunnel and putting the hoops should be done simultaneously otherwise a sudden collapse is very likely. This technique is called "Pish-Kaval" which means putting a hoop as soon as its place would be dug ready. In this method, the worker keeps a hoop on hand while digging the end of the gallery just the size of a hoop, and he installs the hoop immediately after digging. Therefore he has to sit inside the hoops and install them one after another as he digs forward.

Pishkar kani Tunnel extending. Pishkar kani means advancing into the aquifer by digging the end of the gallery in order to increase infiltration area in the tunnel and get more water. This practice is the most common one to increase water discharge of qanat, though it is only possible and allowed where a qanat does not trespass on the bound of another qanat.

Pishkar The tunnel end in water production section is called Pishkar which has the most penetration into the aquifer. The further Pishkar goes into the aquifer, the more water seeps into the tunnel.

Pishvaz Sometimes the workers dig a horizontal tunnel from the half-dug well toward the main gallery that lies at a lower level. Water seepage can no longer hamper the digging of tunnel as it goes parallel to the water table. This tunnel is called "Pishvaz" which means "welcome" for it goes toward the main gallery to welcome it metaphorically.

Pistooyi "Pistooyi" means "off the beaten path" or out of main direction. If two wells would be 50 m away from each other, the tunnel between them is expected to be almost the same 50 m, otherwise the tunnel has gone awry. In this case a tunnel is dug from the second well toward the deviated tunnel in the way that the tunnel goes forward but the well remains apart from this new direction. In Bam region this well is referred to as "Pistooyi".

Pookeh A dry and abandoned qanat. If the water table goes down so much that it is located below the bottom of mother well, no water can seep into the gallery, and if this situation persists, the qanat will inevitably dry up.

Poshte Poshte is the distance between two shaft wells. Shafts are sunk at intervals of 20 to 200 m in a line between the groundwater recharge zone and the irrigated land. This distance is usually twice the depth of wells, though it completely depends on qanat master's judgment and may vary with types of soil and surface topology.

Qanat An almost horizontal tunnel with some shaft wells, which conveys groundwater to the earth surface. This water seeps into the tunnel in the saturated area and then flows down the tunnel toward the earth surface. Qanat drains out water in order to supply potable water and to irrigate the cultivated lands located down slope.

Qanat-e barooni Raining qanat. In plain qanats, water seepage usually takes place in sides of tunnel, whereas in mountainous qanats it is possible to face water seepage even in the ceiling. Such qanats are called "raining qanats" whose tunnel ceiling should be dug like a pitched roof to prevent water drops from falling down directly on the worker's head.

Qanat-e Vaghfi Sometimes some shares of a qanat are devoted to public or religious purposes. This practice is called "Vaqf" whose revenue may be allocated to a mosque, a school, or helping orphans or paupers.

Qomesh Qanat-river, a qanat that conveys the water of river. In this case a surface stream is directed into a qanat nearby through a well which has been constructed for this purpose. To prevent erosion, river water is led to a pool so water settles first and then they let the water enter the qanat. In Dezfool, Khuzestan province qanat system is employed to transfer river water to cultivated lands across some elevated areas in between. These qanats are called "Qomesh" in the region.

Rag-e Zamin A layer of soil. Geological stratification.

Rahrow The qanat gallery which is an almost horizontal tunnel dug to get access to groundwater reserves, and to transfer this water to the earth's surface. The dimensions of the gallery are such that the workers can easily go through and work in it: between 90 and 150 cm high and its width is less than half the height. The length of the tunnel depends on a variety of factors such as the topography, aquifer behavior, qanat discharge, where the water is to be used, etc.

Rassi Rassi is a tool used to avoid turning from the right direction and getting lost underground. Rassi consists of a wooden or iron stick 80–100 cm long with two strings tied to it, some 20 cm away from each other. Two stones hang from the aforesaid strings, forming kind of plumb-bob. The qanat master places the stick horizontally on the mouth of the well A, such that the stick exactly points to the well B, the well to which the tunnel is to go. Now the stick points to the right direction and down the well the stones hanging from the stick point to the same direction. Sometimes the worker holds a lamp just behind the two stones the way that one stone eclipses the other and only one shadow would be shed on the opposite wall. This shadow tells the worker where to dig toward the well B.

Rekhve Soft, damp and crumbling soil in which excavation is difficult.

Rig o Ros Combination of clay and pebble with a good porosity which ease water seepage through it. This soil is suitable for qanat especially in water production section.

Rig-e-Badi Dune which lies beneath younger layers of sediments. This soil is very crumbling and needs to be shored up when qanat cuts through it.

Roo-Neshastan Ascending tunnel. If a qanat worker comes across a hard and rocky layer into which it is difficult to continue digging the tunnel, the worker avoids such a formation by going over it. Going over a hard layer and digging the tunnel with the same slope as that of the layer is called "Roo-Neshastan". In this case both tunnel and hard layer runs parallel toward the mother well.

Ros Ordinary clay which is compact with very low porosity and almost impermeable to water.

Ros-e-Suraxdar Clay with high porosity.

Saat-e Abi Water clock or clepsydra to measure the time every shareholder has for irrigation. Their clepsydra consists of two bowls made of copper one of which is so small that could freely float on the surface of water in the large one. The floating bowl has a tiny hole at its bottom through which water can enter the bowl and gradually fill it up. After being filled which may take a certain time, the small bowl sinks in the water and bumps into the bottom of the large bowl. As soon as the bump would be heard, a unit of time would be over, so the time between the two bumps equals a certain unit of time.

Sandoghi In water transport section, the tunnel ceiling is built angled and rectangular which is called Sandoghi or box-shaped.

Sang Bandi Sang Bandi means lining the qanat tunnel with stones stacked on top of each other with no mortar.

Sang-e-Atishi Flint which is hard and white in color.

Sang-e-rag rag Layered stones which are considered an indication of a good discharge. The qanat masters believe that such stones can convey water from a long distance.

Sanglax Layers of ground awash with peddles and rubble stones with different size and little clay in it, which is widespread in the mountainous area of Yazd.

Sar kooli Refer to Devil.

Sargir A concrete round lid to be put on the well opening to block it.

Sazoo Sazoo is a type of rope made of date palm fibers which is used in shallow wells and on simple windlasses. Privilege of this rope is that it would not be soaked in water and its weight does not change.

Sel Very hard stones regardless of their size, which can be found underground or on the earth surface.

Seqat Rubble stones which lack any cement or clay and can drain out qanat water especially in water transport section.

Shakheh Lateral tunnels. The lateral tunnels are similar to Dastak, branching from the main gallery in different directions. What differentiates these tunnels from one another is their length according to which they have different names. If a lateral tunnel enjoys some shaft wells, it is called "Shakheh" which may be kilometers long.

Shakheh Zani Branching. Shakheh Zani means adding more side branches to the qanat main tunnel which can increase the infiltration area and accordingly the discharge of qanat. In fact each branch acts independently and eventually their waters join and flow down together.

Shaq Sometimes the workers may notice a crack or gap on the tunnel floor which drain out the water flow in water transport section. These cracks or gaps are called Shaq which should be filled up with lime and clay otherwise water escapes through them and the qanat discharge dramatically decreases. If the workers find a Shaq in water production section, they examine it to figure out whether it gives or drains water. If it drains they fill it up with kneaded thick clay mud otherwise they do not manipulate it.

Sheh kani Removing the sediments from the qanat gallery. Sediments should be removed and lifted out every once in a while, otherwise the qanat tunnel gets clogged up. In the mountainous qanats the sediments are denser, harder and more difficult to be broken and removed. The workers do not scrape the sediments off the tunnel walls totally but they try to leave a thin layer of them to keep water percolation at bay in water production section.

Sheh Sediment settled on the qanat sides and floor is called Sheh which should be removed and lifted out every once in a while, otherwise the qanat tunnel gets clogged up.

Shekam Darideh Shekam Darideh literally means "ripped belly" which is a technique in digging a Bizesh. As the worker digs down, the triangle between the gallery, well and Bizesh is removed gradually from top to bottom. That is why in such wells their bottom is much wider than their top, called Shekam Darideh.

Shib-e Rahrow Gradient of qanat gallery. Taking into account the topography and position of the water table, the tunnel gradient is calculated so that the water flowing in the tunnel would have an adequate speed not to leave too much sediment and, on the other hand, would not be so rapid that it washes away the tunnel and causes erosion. The optimum gradient is taken to be between 2/1000 and 5/1000.

Shoolat Very crumbling soil in which it is not possible to tunnel without reinforcing and lining with firm materials.

Shotor galoo Siphon. In case two qanats intersect at the same level, a special structure named "Shotor galoo" is built to pass by the older qanat without digging into its gallery. Shotor galoo literally means "camel's throat", which is actually kind of siphon in which one end is higher than the other, allowing water to come up and flow into the tunnel on the other side due to this height difference.

Sineye Kar The end of the qanat gallery where digging always starts is called Sineye Kar literally "work front". Refer to Pishkar.

Tabari Tabari is a hatchet like pickaxe which has a short and flat end for digging hard soils.

Tahsoo Roosoo Tunnel doubling. If a dramatic drawdown breaks out in the aquifer and groundwater falls over 1.5 m below the qanat tunnel, deepening no longer works and the workers resort to the method of doubling. Tunnel doubling is called Tahsoo Roosoo which means digging another tunnel below the old tunnel and just in parallel with it, so that the new exit point turns up some distance down slope from the old one.

Tanooreh Water house. Tanooreh is a funnel like well with a depth of between 5 and 7 m. It is 1–2 m in diameter at the top and much less at the bottom, with a small outlet at its bottom through which water spouts out and hits the blades of watermill turbine. The operation of the watermill is based on the potential energy of water due to the height of water house: The deeper the water house, the more energy is generated. Sometimes the depth of a water house reaches 10 m below the depth of the qanat, in order to increase the water pressure.

Taraz keshi Leveling. After digging a test well, the next step is to dig water transport section of the gallery, but first of all some calculations are necessary in order to prevent any possible technical mistake. Taraz keshi means calculating the high differences between many spots from the test well down. Thus the qanat practitioners can know how deep each shaft well should be, so that the gallery connecting all the shaft wells would be almost horizontal.

Taraz Leveling tool for calculating the gradient, which consists of two 1.5 m wooden poles with a long thin rope in between. The rope is about 30 m long, tied to the top of both poles. The poles are held upright on the ground in some distance from each other. To make sure that the poles stand completely upright, they cut a rectangular hole in each pole and hang a plummet from the top of hole. The plummet position implies whether or not the pole is completely vertical. To know if the rope between the two poles lies completely horizontal they use a leveling tool which consist of a small glass container full of water with an air bubble in it. They put the glass container gently on the role and consider the rope horizontal if the bubble stays exactly in the middle.

Tare Kar Each qanats consists of two different parts; water production section locally called Tare Kar and water transport section. Part of tunnel dug through the underground saturated area (aquifer) is called water production section. The discharge of the qanat completely depends on the extent, permeability and the depth of this part in comparison to water table. The length of the "tare-Kar" is not fixed but correlates with the fluctuations in water table.

Taroon Refers to Tare Kar.

Tavar Refer to Terek.

Terek Terek is a sound that leads the worker just to the bottom of the half-dug well and the Devil hits the well with no deviation, while digging over head. Terek means a sound heard underground in the tunnel when someone ties a rope around a stone and drops it down the half-dug well and pulls it up and does it again.

Timaj Buckets made of tanned sheep skin.

Towqeh Ceramic or concrete rings to shore up the qanat shaft wells. "Towqeh" is circular in shape with a diameter of 75–80 cm. Those rings are used to prevent collapse only in shaft wells.

Yek Sar Yek Sar or simple windlass is a type of windlass which works only with one rope and one bucket hanging from it.

Zanbil Zanbil is a container used to carry soil, which was made of date fibers. Zanbil was not suitable to carry mud and wet soil because it was torn up easily.

Zangabeh Refer to Sheh.

Zardeh Shoolat Yellow clay which collapses or subsides to a large extent.

Zenama Zenama literally means "Seepage Indicator". Sometimes the owners of qanat take a particular spot as the border between water transport and production sections, whereas the owners of well like to designate a place more upslope as the border so they would have a better chance to exploit more water. Zenama serves to settle the dispute of this kind. A worker who is mutually agreed upon goes down the qanat and makes a cut into the tunnel wall just where believed to be the border by the qanat owners. This cut – Zenama – is some 60–70 cm long and 15 cm deep. The cut is made only 10 cm above the water level flowing down the tunnel. The worker leaves the cut and waits for 24 h, and the day after he returns to the cut to see whether or not any water has accumulated in it. The presence of water in the cut proves that here is not located in the dry zone or water transport section but it is in either water production section or at least the border between.

Zeraa The Zeraa is an ancient unit of length based on the length of the forearm from the elbow to the tip of the middle finger, which equals 45 cm.

Zorna Zorna is a well that directs qanat water to a less elevated level than expected. A typical Zorna may convey water to the earth surface, another shaft well or a qanat gallery. According to where Zorna takes water, it can be classified into three types which are all called Zorna in the qanat workers' jargon.

References

Dehghani A (2009) Water in Iranian Plateau: Qanat, water reservoir and Ice House. Yazda Publication, Tehran

Semsar Yazdi AA (2014) Qanat of Zarch. International Center on Qanats and Historic Hydraulic Structures, Yazd

Semsar Yazdi AA, Labbaf Khaneiki M (2010) Veins of Desert. International Center on Qanats and Historic Hydraulic Structures (ICQHS), Yazd

Bibliography

Aghasi A, Safinejad J (2000) Qanat glossary. Yazd Regional Water Authority, Yazd

Azad M (2000) Artificial recharge of aquifer by controling Qanat water flow, Selected articles on Qanats. Yazd Regional Water Authority, Yazd

Bahadori Nejad M, Dehghani A (2010) Natural and traditional ice production in Iran. Yazda Publication, Yazd

Bazza M (2006) Overview of the history of water resources and irrigation management in the near east region. Food and Agriculture Organization of the United Nations, Cairo
Bonine ME (1982) From qanat to kort: traditional irrigation terminology and practices in central Iran. IRAN J Br Inst Persian Stud XX:145–159
Boucharlat R (2001) Iron age water-draining galleries and the Iranian Qanat. In: Proceedings of the first international conference on the archeology of the U.A.E
Briant P (2001) Irrigation et drainage dans l'Antiquité, qanâts et canalisations souterraines en Iran, en Égypte et en Grèce, Paris, Thotm Éditions (Persika 2), 190 p., ill
Chauveau M (2001) Les qanates dans les ostraca de Manawir, Paris, Seminaire tenu au au colloge de France, Persika 2
Cheraghi Z (2010) Role of Qanats in the historic life of Yazd from Ilkhanid Dynasty to Pahlavi I Era (1295–1941); a critical study. Isfahan University
Dehghani A (2009) Water in Iranian Plateau: Qanat, water reservoir and Ice House. Yazda Publication, Tehran
Diamond J (2005) Collapse: how societies choose to fail or survive. Penguin Group, Camberwell
Ellis AJ (1917) The divining rod; a history of water witching, United States Geological Survey, Department of the Interior. Government Printing Office, Washington
Garbrecht G (1986) Wasserspeicher (Talsperren) in der Antike, Antike Welt, 2nd special edition: Antiker Wasserbau
Goblot H (1979) Les Qanats, une technique d'acquisition de l'eau/English: Qanat a technique for obtaining water, Paris-La Haye, Mouton/Ecole des hautes en sciences sociales, 236 P. Translated from French to Persian by A. Sarvqad Moqadam, M. H. Papoli Yazdi, 1992
Hasanalian D (2006) Sialk; the unsolved puzzle of Archaeology. Iran Newspaper Archive at http://www.magiran.com/npview.asp?ID=1295401
Haupt L (1925) Armenien einst und yetst. Leipzig
Honari M, Askari H, Khosrowchahli M (2014) Use of desirability function method in optimization of regeneration and Callus induction of Alhagi camelorum. American Journal of Plant Sciences 5:268–274
Jacobsen T, Lloyd S (1935) Sennacherib's Aqueduct at Jerwan, Oriental Institute Publication 24. University of Chicago Press, Chicago
Karaji Abu Bakr ibn Muhammad ibn al Hassan (1966) Extraction of hidden waters, Translated from Arabic to Persian by Hossein Khadiv Jam. Bonyad Farhang Iran Publication, Tehran
Kobori I (1964) Some considerations on the origin of the Qanat system. In: Memorial collected papers dedicated to Prof. E. Ishida. Tokyo
Labbaf Khaneiki M (2009) Final report of Shekh Mamudian Kahriz renovation. International Center on Qanats & Historic Hydraulic Structures (ICQHS) under the auspices of UNESCO, unpublished
Labbaf Khaneiki M (2015) Qanat: a response to the climate change in Iranian Plateau. Asar Sci J, Research Center of Iranian Cultural Heritage Organization, No. 66
Lightfoot DR (1996) Syrian Qanat Romani: history, ecology, abandonment. Journal of Arid Environments 33(3):321–336
Lightfoot DR (1997) Qanats in the Levant: hydraulic technology at the periphery of early empires. Technology and Culture 38(2):432–451
Narshakhi A (943 AD). Khanji AH (ed).Electronic Publication of Iranian History
Papoli Yazdi MH, Labbaf Khaneiki M (2000) The role of Qanat in formation of civilizations; sustainability of Qanat civilization and culture. In: Proceedings of the international conference on Qanat, vol. 1, Yazd Regional Water Authority, Yazd
Papoli Yazdi MH, Labbaf Khaneiki M, Labbaf Khaneiki R, Jalali A (2000) Qanat of Ghasabeh Gonabad as a myth. Khorasan Regional Water Authority, Bojnord
Pollock S, Bernbeck R (2005) Archaeologies of the Middle East critical perspectives. Blackwell Publishing Ltd, Malden
Rashed R (1970–1980), Al-Karajī Abū Bakr Ibn Muḥammad Ibn al Ḥusayn. Dictionary of scientific biography. Charles Scribner's Sons, New York

Safinejad J (2000a) Desert and Mountanous Qanats in Kashan, Selected articles on Qanats. Yazd Regional Water Authority, Yazd
Safinejad J (2000b) Two Storey Qanat of Ardestan, Book of selected articles on Qanat. Yazd Regional Water Authorities, Yazd (in Persian)
Safinejad J, Dadras B (2000) Underground Dam of Vazvan Qanat in Isfahan. Iranian National Water Museum, Tehran
Salmasi F (1997) Application of computer modeling in the management of artificial groundwater recharge in the Plain of Ghorveh. University of Ahvaz Shahid Chamran, Ahvaz
Salvini M (2001) Pas de Qanat en Urartu, Paris, Irrigation et drainage dans l'antiquite, qanats et canalisation souteraines en Iran, Egypt et en Grece
Semsar Yazdi AA (2010) Qanat from practitioners' point of view. Iran Water Resources Management Company, Tehran
Semsar Yazdi AA (2014) Qanat of Zarch. International Center on Qanats and Historic Hydraulic Structures, Yazd
Semsar Yazdi AA, Labbaf Khaneiki M (2010) Veins of desert. International Center on Qanats and Historic Hydraulic Structures (ICQHS), Yazd
Semsar Yazdi AA, Labbaf Khaneiki M (2011) Qanat in its Cradle, volume 1. International Center on Qanats and Historic Hydraulic Structures (ICQHS), Yazd
Weisgerber G (2003) The impact of the dynamics of Qanats and Aflaj on oases in Oman: comparisons with Iran and Bahrain. Internationales frontinus-symposium, Luxemburg
Wittfogel KA (1957) Oriental despotism: a comparative study of total power. Yale University Press, London
Worster D (1985) Rivers of empire: water, aridity, and the growth of the American West, 1st edn. Pantheon Books, New York

Index

A.A. Semsar Yazdi, M. Labbaf Khaneiki, *Qanat Knowledge*,
DOI 10.1007/978-94-024-0957-4

MIX
Papier aus verantwortungsvollen Quellen
Paper from responsible sources
FSC® C105338

FSC
www.fsc.org

If you have any concerns about our products,
you can contact us on
ProductSafety@springernature.com

In case Publisher is established outside the EU,
the EU authorized representative is:
Springer Nature Customer Service Center GmbH
Europaplatz 3, 69115 Heidelberg, Germany

Printed by Libri Plureos GmbH
in Hamburg, Germany